# SUMMING IT UP

# SUMMING IT UP

## From One Plus One to Modern Number Theory

**AVNER ASH** AND
**ROBERT GROSS**

PRINCETON UNIVERSITY PRESS
PRINCETON AND OXFORD

Library of Congress Cataloging-in-Publication Data

Names: Ash, Avner, 1949– | Gross, Robert, 1959–
Title: Summing it up : from one plus one to modern number theory /
Avner Ash and Robert Gross.
Description: Princeton : Princeton University Press, [2016] | Includes bibliographical
references and index.
Identifiers: LCCN 2015037578 | ISBN 9780691170190 (hardcover)
Subjects: LCSH: Number theory. | Mathematics—Popular works.
Classification: LCC QA241 .A85 2016 | DDC 512.7—dc23 LC
record available at http://lccn.loc.gov/2015037578

British Library Cataloging-in-Publication Data is available

This book has been composed in New Century Schoolbook

Printed on acid-free paper. ∞

Typeset by S R Nova Pvt Ltd, Bangalore, India
Printed in the United States of America

1 3 5 7 9 10 8 6 4 2

*For our families*

"Manners are not taught in lessons," said Alice. "Lessons teach
you to do sums, and things of that sort."
"And you do Addition?" the White Queen asked. "What's one
and one and one and one and one and one and one and
one and one and one?"
"I don't know," said Alice. "I lost count."
"She can't do Addition," the Red Queen interrupted.
—Lewis Carroll, *Through the Looking-Glass*


Numbers it is. All music when you come to think. Two multiplied
by two divided by half is twice one. Vibrations: chords those are.
One plus two plus six is seven. Do anything you like with
figures juggling. Always find out this equal to that. Symmetry
under a cemetery wall.
—James Joyce, *Ulysses*


... my true love is grown to such excess
I cannot sum up sum of half my wealth.
—William Shakespeare, *Romeo and Juliet*, II.vi.33–34

# Contents

*Preface*                                                                   xi

*Acknowledgments*                                                           xv


INTRODUCTION: WHAT THIS BOOK IS ABOUT                                         1
  1. Plus                                                                     1
  2. Sums of Interest                                                         3


PART ONE.  FINITE SUMS

CHAPTER 1.  PROEM                                                            11
  1. Greatest Common Divisors                                                11
  2. Congruences                                                             14
  3. Wilson's Theorem                                                        15
  4. Quadratic Residues and Nonresidues                                      17
  5. The Legendre Symbol                                                     19

CHAPTER 2.  SUMS OF TWO SQUARES                                             22
  1. The Answer                                                              22
  2. The Proof Is Not in the Pudding                                         26
  3. The "If" Parts of Theorems 2.1 and 2.3                                  28
  4. The Details                                                             29

CHAPTER 3.  SUMS OF THREE AND FOUR SQUARES                                  32
  1. Three Squares                                                           32
  2. Interlude                                                               33

3. Four Squares                                              34
4. Sums of More Than Four Squares                            35

CHAPTER 4. SUMS OF HIGHER POWERS: WARING'S
PROBLEM                                                      37
1. $g(k)$ and $G(k)$                                         37
2. Sums of Biquadrates                                       39
3. Higher Powers                                             40

CHAPTER 5. SIMPLE SUMS                                       42
1. Return to First Grade                                     42
2. Adding Small Powers                                       43

CHAPTER 6. SUMS OF POWERS, USING LOTS
OF ALGEBRA                                                   50
1. History                                                   50
2. Squares                                                   52
3. Divertimento: Double Sums                                 55
4. Telescoping Sums                                          57
5. Telescoping Sums Redux                                    59
6. Digression: Euler–Maclaurin Summation                     66

PART TWO. INFINITE SUMS

CHAPTER 7. INFINITE SERIES                                   73
1. Finite Geometric Series                                   73
2. Infinite Geometric Series                                 75
3. The Binomial Series                                       76
4. Complex Numbers and Functions                             79
5. Infinite Geometric Series Again                           81
6. Examples of Infinite Sums                                 83
7. $e$, $e^x$, and $e^z$                                     85
8. Power Series                                              87
9. Analytic Continuation                                     91

CHAPTER 8. CAST OF CHARACTERS 96
 1. $H$ 96
 2. $e^z$ Again 97
 3. $q$, $\Delta^*$, and $\Delta^0$ 98

CHAPTER 9. ZETA AND BERNOULLI 103
 1. A Mysterious Formula 103
 2. An Infinite Product 104
 3. Logarithmic Differentiation 106
 4. Two More Trails to Follow 109

CHAPTER 10. COUNT THE WAYS 110
 1. Generating Functions 110
 2. Examples of Generating Functions 113
 3. Last Example of a Generating Function 119


PART III. MODULAR FORMS AND THEIR APPLICATIONS

CHAPTER 11. THE UPPER HALF-PLANE 127
 1. Review 127
 2. The Strip 128
 3. What Is a Geometry? 130
 4. Non-Euclidean Geometry 132
 5. Groups 134
 6. Matrix Groups 138
 7. The Group of Motions of the Hyperbolic
     Non-Euclidean Plane 141

CHAPTER 12. MODULAR FORMS 147
 1. Terminology 147
 2. $SL_2(\mathbf{Z})$ 148
 3. Fundamental Domains 150
 4. Modular Forms at Last 153
 5. Transformation Property 155
 6. The Growth Condition 158
 7. Summary 158

CHAPTER 13. HOW MANY MODULAR FORMS ARE
THERE?                                                                    160
1. How to Count Infinite Sets                                            160
2. How Big Are $M_k$ and $S_k$?                                          164
3. The $q$-expansion                                                     169
4. Multiplying Modular Forms                                             171
5. Dimensions of $M_k$ and $S_k$                                         175

CHAPTER 14. CONGRUENCE GROUPS                                            179
1. Other Weights                                                         179
2. Modular Forms of Integral Weight and Higher Level                     182
3. Fundamental Domains and Cusps                                         182
4. Modular Forms of Half-Integral Weight                                 184

CHAPTER 15. PARTITIONS AND SUMS OF SQUARES
REVISITED                                                                186
1. Partitions                                                            186
2. Sums of Squares                                                       190
3. Numerical Example and Philosophical Reflection                        196

CHAPTER 16. MORE THEORY OF MODULAR FORMS                                 201
1. Hecke Operators                                                       201
2. New Clothes, Old Clothes                                              208
3. $L$-functions                                                         210

CHAPTER 17. MORE THINGS TO DO WITH MODULAR
FORMS: APPLICATIONS                                                      213
1. Galois Representations                                                214
2. Elliptic Curves                                                       217
3. Moonshine                                                             219
4. Bigger Groups (Sato–Tate)                                             221
5. Envoy                                                                 223

Bibliography                                                             225

Index                                                                    227

# Preface

Adding two whole numbers together is one of the first things we learn in mathematics. Addition is a rather simple thing to do, but it almost immediately raises all kinds of curious questions in the mind of an inquisitive person who has an inclination for numbers. Some of these questions are listed in the Introduction on page 3. The first purpose of this book is to explore in a leisurely way these and related questions and the theorems they give rise to. You may read a more detailed discussion of our subject matter in the Introduction.

It is in the nature of mathematics to be precise; a lack of precision can lead to confusion. For example, in Bloom's musings from *Ulysses*, quoted as one of our epigraphs, a failure to use exact language and provide full context makes it appear that his arithmetical claims are nonsense. They may be explained, however, as follows: "Two multiplied by two divided by half is twice one" means $\frac{2+2}{2} = 2 \cdot 1$, because Bloom meant "cut in half" when he used the ambiguous phrase "divided by half." Then "$1 + 2 + 6 = 7$" refers to musical intervals: If you add a unison, a second, and a sixth, the result is indeed a seventh. Bloom himself notes that this can appear to be "juggling" if you suppress a clear indication of what you are doing.

Bloom has the luxury of talking to himself. In this book, we strive to be clear and precise without being overly pedantic. The reader will decide to what extent we have succeeded. In addition to clarity and precision, rigorously logical proofs are characteristic of mathematics. All of the mathematical assertions in this book can be proved, but the proofs often are too intricate for us to discuss in any detail. In a textbook or research monograph, all such proofs would be given, or reference made to places where they could be found. In a book such as this, the reader must trust us that all of our mathematical assertions have proofs that have been verified.

This book is the third in a series of books about number theory written for a general mathematically literate audience. (We address later exactly what we mean by "mathematically literate.") The first two books were *Fearless Symmetry* and *Elliptic Tales* (Ash and Gross, 2006; 2012). The first book discussed problems in Diophantine equations, such as Fermat's Last Theorem (FLT). The second discussed problems related to elliptic curves, such as the Birch–Swinnerton-Dyer Conjecture. In both of these books, we ended up mumbling something about modular forms, an advanced topic that plays a crucial role in both of these areas of number theory. By the time we reached the last chapters in these books, we had already introduced so many concepts that we could only allude to the theory of modular forms. One purpose of *Summing It Up* is to give in Part III a more leisurely and detailed explanation of modular forms, motivated by the kinds of problems we will discuss in Parts I and II.

Each of the three books in our trilogy may be read independently of the others. After reading the first two parts of *Summing It Up*, a very diligent person might gain from reading *Fearless Symmetry* or *Elliptic Tales* in tandem with the third part of *Summing It Up*, for they provide additional motivation for learning about modular forms, which are dealt with at length in Part III. Of course, this is not necessary—we believe that the number-theoretical problems studied in the first two parts by themselves lead naturally to a well-motivated study of modular forms.

The three parts of this book are designed for readers of varying degrees of mathematical background. Part I requires a knowledge of high school algebra and geometry. Only in a few dispensable sections is any deeper mathematical knowledge needed. Some of the exposition involves complicated and sometimes lengthy strings of algebraic manipulation, which can be skipped whenever you wish. To read Part II, you will need to have encountered much of the content of the first year of a standard calculus course (mostly infinite series, differentiation, and Taylor series). You will also need to know about complex numbers, and we review them briefly. Part III does not require any additional mathematical knowledge, but it gets rather intricate. You may need a good dose of patience to read through all of the details.

The level of difficulty of the various chapters and sections sometimes fluctuates considerably. You are invited to browse them in any order. You can always refer back to a chapter or section you skipped, if necessary, to fill in the details. However, in Part III, things will probably make the most sense if you read the chapters in order.

It continues to amaze us what human beings have accomplished, starting with one plus one equals two, getting to two plus two equals four (the cliché example of a simple truth that we know for sure is true), and going far beyond into realms of number theory that even now are active areas of research. We hope you will enjoy our attempts to display some of these wonderful ideas in the pages that follow.

# Acknowledgments

We wish to thank the anonymous readers employed by Princeton University Press, all of whom gave us very helpful suggestions for improving the text of this book. Thanks to Ken Ono and David Rohrlich for mathematical help, Richard Velkley for philosophical help, and Betsy Blumenthal for editorial help. Thanks to Carmina Alvarez and Karen Carter for designing and producing our book. Great thanks, as always, to our editor, Vickie Kearn, for her unfailing encouragement.

# SUMMING IT UP

# WHAT THIS BOOK IS ABOUT

## 1. Plus

Counting—one, two, three, four or *uno, dos, tres, cuatro* (or in whatever language); or I, II, III, IV or 1, 2, 3, 4, or in whatever symbols—is probably the first theoretical mathematical activity of human beings. It is theoretical because it is detached from the objects, whatever they might be, that are being counted. The shepherd who first piled up pebbles, one for each sheep let out to graze, and then tossed them one by one as the sheep came back to the fold, was performing a practical mathematical act—creating a one-to-one correspondence. But this act was merely practical, without any theory to go with it.

This book is concerned with what may have been the next mathematical activity to be discovered (or invented): addition. We may think that addition is primitive or easy, because we teach it to young children. A moment's reflection will convince you that it must have taken an enormous intellectual effort to conceive of an abstract theory of addition. There cannot be addition of two numbers before there are numbers, and the formation of pure numbers is sophisticated because it involves abstraction.

You may want to plunge into the mathematics of addition and skip the rest of this section. Or you may want to muse a little on philosophical problems connected with the concepts and practice of mathematics, as exemplified in the subject matter of this book. Those who have a taste for these philosophical questions may enjoy the following extremely brief survey.

Perhaps this abstraction of pure numbers was formed through the experience of counting. Once you had a series of number words, you could count axes one day, sheep the next, and apples the third day. After a while, you might just recite those words without any particular thing being counted, and then you might stumble onto the concept of a pure number. It is more probable that arithmetic and the abstract concept of number were developed together.[1]

One way to see the difficulty of the concepts of number, counting, and addition is to look at the philosophy of mathematics, which to this day has not been able to decide on a universally acceptable definition of "number." The ancient Greek philosophers didn't even consider that the number one was a number, since in their opinion numbers were what we counted, and no one would bother to count "one, period."

We won't say more about the very difficult philosophy of mathematics, except to mention one type of issue. Immanuel Kant and his followers were very concerned with addition and how its operations could be justified philosophically. Kant claimed that there were "synthetic truths *a priori*." These were statements that were true, and could be known by us to be true prior to any possible experience, but whose truth was not dependent on the mere meaning of words. For example, the statement "A bachelor doesn't have a wife" is true without needing any experience to vouch for its truth, because it is part of the definition of "bachelor" that he doesn't have a wife. Such a truth was called "analytic *a priori*." Kant claimed that "five plus seven equals twelve" was indubitably true, not needing any experience to validate its truth, but it was "synthetic" because (Kant claimed) the concept of "twelve" was not logically bound up and implied by the concepts of "five," "seven," and "plus." In this way, Kant could point to arithmetic to show that synthetic truths *a priori* existed, and then could go on to consider other such truths that came up later in his philosophy.

In contrast, other philosophers, such as Bertrand Russell, have thought that mathematical truths are all analytic.

---

[1] Of course, in practice, two apples are easily added to two apples to give four apples. But a theoretical approach that could lead to the development of number theory is more difficult. There was a period in the mathematical thought of the ancient Greek world when "pure" numbers were not clearly distinguished from "object" numbers. For an account of the early history of arithmetic and algebra, we recommend Klein (1992).

These philosophers often think logic is prior to mathematics. And then there is the view that mathematical truths are "*a posteriori*," meaning they are dependent on experience. This seems to have been Ludwig Wittgenstein's opinion. It was also apparently the opinion of the rulers in George Orwell's novel *1984*, who were able to get the hero, in defeat, to believe firmly that two plus two equals five.

The philosophy of mathematics is exceedingly complicated, technical, and difficult. During the twentieth century, things grew ever more vexed. W.V.O. Quine questioned the analytic–synthetic distinction entirely. The concept of truth (which has always been a hard one to corner) became more and more problematic. Today, philosophers do not agree on much, if anything, concerning the philosophical foundations of numbers and their properties. Luckily, we do not need to decide on these philosophical matters to enjoy some of the beautiful theories about numbers that have been developed by mathematicians. We all have some intuitive grasp of what a number is, and that grasp seems to be enough to develop concepts that are both free from contradictions and yield significant theorems about numbers. We can test these theorems doing arithmetic by hand or by computer. By verifying particular number facts, we have the satisfaction of seeing that the theorems "work."

## 2. Sums of Interest

This book is divided into three parts. The first part will require you to know college algebra and Cartesian coordinates but, except in a few places, nothing substantially beyond that. In this part, we will ask such questions as:

- Is there a short formula for the sum $1 + 2 + 3 + \cdots + k$?
- How about the sum $1^2 + 2^2 + 3^2 + \cdots + k^2$?
- We can be even more ambitious. Let $n$ be an arbitrary integer, and ask for a short formula for
  $1^n + 2^n + 3^n + \cdots + k^n$.

- How about the sum $1 + a + a^2 + \cdots + a^k$?
- Can a given integer $N$ be written as a sum of perfect squares? Cubes? $n$th powers? Triangular numbers? Pentagonal numbers?
- Obviously, an integer larger than 1 can be written as a sum of smaller positive integers. We can ask: In how many different ways can this be done?
- If a number can be written as a sum of $k$ squares, in how many different ways can this be done?

Why do we ask these questions? Because it is fun and historically motivated, and the answers lead to beautiful methods of inquiry and amazing proofs.

In the second part of this book, you will need to know some calculus. We will look at "infinite series." These are infinitely long sums that can only be defined using the concept of *limit*. For example,

$$1 + 2 + 3 + \cdots = ?$$

Here the dots mean that we intend the summing to go on "forever." It seems pretty clear that there can't be any answer to this sum, because the partial totals just keep getting bigger and bigger. If we want, we can define the sum to be "infinity," but this is just a shorter way of saying what we said in the previous sentence.

$$1 + 1 + 1 + \cdots = ?$$

This, too, is clearly infinity.

How should we evaluate

$$1 - 1 + 1 - 1 + 1 - 1 + 1 - \cdots = ?$$

Now you might hesitate. Euler[2] said this sum added up to $\frac{1}{2}$.

$$1 + a + a^2 + \cdots = ?$$

---

[2] One way to justify Euler's answer is to use the formula for the sum of an infinite geometric series. In chapter 7, section 5, we have the formula

$$\frac{1}{1-z} = 1 + z + z^2 + z^3 + \cdots,$$

valid as long as $|z| < 1$. If we dare to go outside the region of validity and substitute $z = -1$, we see why Euler interpreted the sum as he did.

We will see that this problem has a nice answer if $a$ is a real number strictly between $-1$ and $1$. You may have learned this answer when you studied "geometric series." We will expand our algebra so that we can use a complex number for $a$.

Then we can ask about

$$1^n + 2^n + 3^n + \cdots = ?$$

where $n$ is any complex number. This answer (for some values of $n$) gives a function of $n$ called the $\zeta$-function.

Going back one step, we can add in coefficients:

$$b_0 + b_1 a + b_2 a^2 + \cdots = ?$$

This is the setting in which we introduce the concept of *generating function*, where $a$ itself is a variable.

We can also add coefficients to the $\zeta$-function series and ask about series like

$$c_1 1^n + c_2 2^n + c_3 3^n + \cdots = ?$$

which are called *Dirichlet series*.

These questions and their answers put us in position in the third part of this book to define and discuss *modular forms*. The surprising thing will be how the modular forms tie together the subject matter of the first two parts. This third part will require a little bit of group theory and some geometry, and will be on a somewhat higher level of sophistication than the preceding parts.

One motivation for this book is to explain modular forms, which have become indispensable in modern number theory. In both of our previous two books, modular forms appeared briefly but critically toward the end. In this book, we want to take our time and explain some things about them, although we will only be scratching the surface of a very broad and deep subject. At the end of the book, we will review how modular forms were used in Ash and Gross (2006) to tie up Galois representations and prove Fermat's Last Theorem, and in Ash and Gross (2012) to be able to phrase the fascinating Birch–Swinnerton-Dyer Conjecture about solutions to cubic equations.

As a leitmotif for this book, we will take the problem of "sums of squares," because it is a very old and very pretty problem whose solution is best understood through the theory of modular forms. We can describe the problem a little now.

Consider a whole number $n$. We say $n$ is a *square* if it equals $m^2$, where $m$ is also a whole number. For example, 64 is a square, because it is 8 times 8, but 63 is not a square. Notice that we define $0 = 0^2$ as a square, and similarly $1 = 1^2$. We easily list all squares by starting with the list 0, 1, 2,... and squaring each number in turn. (Because a negative number squared is the same as its absolute value squared, we only have to use the nonnegative integers.) We obtain the list of squares

$$0, 1, 4, 9, 16, 25, 36, 49, 64, 81, 100, \ldots,$$

in which, as you can see, the squares get farther and farther apart as you go along. (PROOF: The distance between successive squares is $(m + 1)^2 - m^2 = m^2 + 2m + 1 - m^2 = 2m + 1$, so the distance gets larger as $m$ gets larger. Notice that this argument gives us more precise information: The list of differences between successive squares is in fact the list of positive odd numbers in increasing order.) We can be very pedantic, if a bit ungrammatical, and say that the list of squares is the list of numbers that are "the sum of one square."

This raises a more interesting question: What is the list of numbers that are the sum of two squares? You could write a computer program that would output this list up to some limit $N$, and your computer program could generate the list in at least two different ways. First, list all the squares up to $N$. Then:

METHOD 1: Add pairs of squares on your list together in all possible ways. Then arrange the answers in ascending order.

METHOD 2: Form a loop with $n$ going from 0 to $N$. For each $n$, add up all pairs of squares less than or equal to $n$ to see if you get $n$. If you do, put $n$ on the list and go on to $n + 1$. If you don't, leave $n$ off the list, and go on to $n + 1$.

NOTE: We have defined 0 as a square, so any square number is also a sum of two squares. For example, $81 = 0^2 + 9^2$. Also, we allow a square to be reused, so twice any square number is a sum of two squares. For example, $162 = 9^2 + 9^2$.

Run your program, or add squares by hand. Either way, you get a list of the sums of two squares that starts out like this:

$$0, 1, 2, 4, 5, 8, 9, 10, 13, \dots.$$

As you can see, not every number is on the list, and it is not immediately clear how to predict if a given number will be a sum of two squares or not. For instance, is there a way to tell if 12345678987654321 is on the list without simply running your computer program? Nowadays, your program would probably only take a fraction of a second to add up all squares up to 12345678987654321, but we can easily write down a number large enough to slow down the computer. More importantly, we would like a theoretical answer to our question, whose proof would give us some insight into which numbers are on the list and which ones are not.

Pierre de Fermat asked this question in the seventeenth century[3] and must have made such a list. There were no computers in the seventeenth century, so his list could not have been all that long, but he was able to guess the correct answer as to which numbers are sums of two squares.[4] In chapter 2, we will supply the answer and discuss the proof in a sketchy way. Because this book is not a textbook, we don't want to give complete proofs. We prefer to tell a story that may be more easily read. If you wish, you can refer to our references and find the complete proof.

Once you get interested in this kind of problem (as did Fermat, who gave a huge impulse to the study of number theory), then it is easy to create more of them. Which numbers are sums of three squares? Of four squares? Of five squares? This particular list of puzzles will stop because, 0 being a square, any sum of four squares

---

[3] Apparently, another mathematician, Albert Girard, had asked the question and guessed the answer before Fermat, but Fermat publicized the problem.

[4] Fermat announced the answer in a letter to another mathematician, Marin Mersenne, without giving a proof. The first printed proof was written by Leonhard Euler.

will also be a sum of five, six, or any larger amount of squares, and we will see that in fact every positive integer is a sum of four squares.

You could also ask: Which numbers are sums of two cubes, three cubes, four cubes, and so on? You could then substitute higher powers for cubes.

You could ask, as did Euler: Any number is a sum of four squares. A geometrical square has four sides. Is every number a sum of three triangular numbers, five pentagonal numbers, and so on? Cauchy proved that the answer is "yes."

At some point in mathematical history, something very creative happened. Mathematicians started to ask an apparently harder question. Rather than wonder if $n$ can be written as a sum of 24 squares (for example), we ask: *In how many different ways* can $n$ be written as the sum of 24 squares? If the number of ways is 0, then $n$ is not a sum of 24 squares. But if $n$ is a sum of 24 squares, we get more information than just a yes/no answer. It turned out that this harder question led to the discovery of powerful tools that have a beauty and importance that transcends the puzzle of sums of powers, namely tools in the theory of generating functions and modular forms. And that's another thing that this book is about.

# PART ONE

# Finite Sums

# PROEM

In the interest of allowing the reader to enjoy our book without constantly referring to many other references, we collect in this chapter many standard facts that we will often use in the remainder of the book. A reader familiar with elementary number theory can skip this chapter and refer back to it when necessary. We covered most of these topics in Ash and Gross (2006).

## 1. Greatest Common Divisors

If $a$ is a positive integer and $b$ is any integer, then long division tells us that we can always divide $a$ into $b$ and get an integer quotient $q$ and integer remainder $r$. This means that $b = qa + r$, and the remainder $r$ always satisfies the inequality $0 \leq r < a$. For example, if we take $a = 3$ and $b = 14$, then $14 = 4 \cdot 3 + 2$; the quotient $q = 4$ and the remainder $r = 2$. You may not be used to thinking about it, but you can do this with $b < 0$ also. Take $b = -14$ and $a = 3$, and $-14 = (-5) \cdot 3 + 1$; the quotient is $q = -5$, and the remainder is $r = 1$. Notice that if we divide by 2, the remainder will always be 0 or 1; if we divide by 3, the remainder will always be 0, 1, or 2; and so on.

If the result of the long division has $r = 0$, then we say that "$a$ divides $b$." We write this sentence symbolically as $a \mid b$. Of course, one requirement for long division is that $a$ cannot be 0, so whenever we write $a \mid b$, we implicitly assert that $a \neq 0$. If the remainder $r$ is not zero, we say that "$a$ does not divide $b$." We write that assertion symbolically as $a \nmid b$. For example, $3 \mid 6$, $3 \nmid 14$, and $3 \nmid (-14)$. Notice that if $n$ is any integer (even 0), then $1 \mid n$. Also, if $a$ is any positive integer, then $a \mid 0$. At the risk of giving too many examples, we also point out that $2 \mid n$ means that $n$ is even, and $2 \nmid n$ means that $n$ is odd.

Suppose now that $m$ and $n$ are integers that are not both 0. We can then define the greatest common divisor:

**DEFINITION**: The *greatest common divisor* of $m$ and $n$, symbolically written $(m, n)$, is the largest integer $d$ such that $d \mid m$ and $d \mid n$. If the greatest common divisor of $m$ and $n$ is 1, we say that $m$ and $n$ are *relatively prime*.

Because all divisors of $m$ are at most as big as $m$ (if $m > 0$) or $-m$ (if $m < 0$), we can theoretically list all divisors of $m$ and all divisors of $n$, and then pick the largest number that is on both lists. We know that the number 1 is on both lists, and there may or may not be any larger number simultaneously on both lists. For example, $(3, 6) = 3$, $(4, 7) = 1$, $(6, 16) = 2$, and $(31, 31) = 31$. This process would be tedious, though, if we wanted to compute $(1234567, 87654321)$. There is a process called the *Euclidean algorithm*, which allows one to compute greatest common divisors without listing all of the divisors of both $m$ and $n$. We will not describe that process here, but we will state and prove one consequence, often called *Bézout's identity*.

**THEOREM 1.1**: *Suppose that $m$ and $n$ are not both 0, and suppose that $d$ is the greatest common divisor of $m$ and $n$. Then there are integers $\lambda$ and $\mu$ such that $d = \lambda m + \mu n$.*

You can skip the proof if you like. It's actually a frustratingly incomplete proof, because we aren't going to tell you how to find $\lambda$ and $\mu$. Part of what the Euclidean algorithm does is to let you find $\lambda$ and $\mu$ quickly.

**PROOF**: Let $S$ be the following very complicated set, in which the symbol $\mathbf{Z}$ stands for the set of all integers:

$$S = \{am + bn \mid a, b \in \mathbf{Z}\}.$$

In words, $S$ is the set of all multiples of $m$ (positive, negative, and 0) added to all multiples of $n$ (ditto). Because $S$ contains $0 \cdot m + 0 \cdot n$, we know that $S$ contains 0. Because $S$ contains $m$, $-m$, $n$, and $-n$, we know that $S$ contains some positive integers and some negative integers, whether $m$ and $n$ are positive or negative. Moreover, if we add two numbers in $S$, we get a number that is in $S$.[1] One more nonobvious assertion is that if $s$ is any number in $S$, then every multiple of $s$ is also in $S$.[2]

Now, find the integer $d$ that is the smallest positive integer in $S$. (Here's where we are using a quite subtle fact: If $T$ is any set of integers that contains some positive integers, then there is some number that is the smallest positive integer in $T$.) We know that $d$ is the sum of a multiple of $m$ and a multiple of $n$, so write $d = \lambda m + \mu n$. We are now going to prove three assertions:

(1) $d \mid m$.

(2) $d \mid n$.

(3) If $c \mid m$ and $c \mid n$, then $c \leq d$.

After we prove these assertions, we can conclude that $d$ is the greatest common divisor of $m$ and $n$.

Let's try dividing $m$ by $d$. We know that we can write $m = qd + r$, where $0 \leq r < d$. Let's rewrite that equation as $r = (-q)d + m$. We know that $m$ is an element of $S$ because it's $1 \cdot m + 0 \cdot n$. We know that $d$ is an element of $S$, and therefore every multiple of $d$ is an element of $S$. In particular, $(-q)d$ is an element of $S$. We know that when we add two elements of $S$, we always get an element of $S$. Therefore, we are sure that $r$ is an element of $S$.

But $r$ is smaller than $d$, and we picked $d$ to be the smallest positive element in $S$. We are forced to conclude that $r = 0$, which, at long last, tells us that $d$ divides $m$. A similar argument shows that $d$ divides $n$.

---

[1] PROOF: $(a_1 m + b_1 n) + (a_2 m + b_2 n) = (a_1 + a_2)m + (b_1 + b_2)n$.
[2] PROOF: If $s = a_1 m + b_1 n$, then $ks = (ka_1)m + (kb_1)n$.

Now we know that $d$ is a common divisor of both $m$ and $n$. How do we know that $d$ is the largest number that divides both $m$ and $n$? Suppose that $c$ is a positive integer that divides both $m$ and $n$. We can write $m = q_1 c$ and $n = q_2 c$. We know that $d = \lambda m + \mu n$ for some integers $\lambda$ and $\mu$, because $d$ is an element of $S$. Substitution tells us that $d = c(\lambda q_1 + \mu q_2)$. In other words, $c$ divides $d$, so $c$ cannot be larger than $d$. So $d$ is the greatest common divisor of $m$ and $n$, and $d = \lambda m + \mu n$, as we just said.  $\square$

NOTE: We do not give full proofs of many theorems in this book. When we do give a proof, as we just did, the end of the proof is marked with a square $\square$.

One of many consequences of theorem 1.1 is referred to both as the Fundamental Theorem of Arithmetic and as Unique Prime Factorization. Remember a basic definition:

**DEFINITION**: A *prime* is a number $p$ that is larger than 1 and has no positive divisors other than 1 and $p$.

**THEOREM 1.2**: *Suppose that $n$ is an integer that is larger than 1. Then there is one and only one way to factor $n$ into primes:*

$$n = p_1^{e_1} p_2^{e_2} \cdots p_k^{e_k},$$

*where each $p_i$ is prime, $p_1 < p_2 < \cdots < p_k$, and each $e_i > 0$.*

The reason that our phrasing is so detailed is that there may be many ways to factor some numbers into a product of primes: $12 = 2 \cdot 2 \cdot 3$, $12 = 2 \cdot 3 \cdot 2$, and $12 = 3 \cdot 2 \cdot 2$, for example. But these are really all the same factorization, once we restrict our product formula to list the primes in increasing order.

## 2. Congruences

Suppose that $n$ is an integer that is larger than 1. We write $a \equiv b$ (mod $n$), read in words as "$a$ is congruent to $b$ modulo $n$," for the assertion that $n \mid (a - b)$. The number $n$ is called the *modulus* of the congruence. Congruence is an *equivalence relation*, which means that for fixed $n$:

(C1) $a \equiv a$ (mod $n$).

(C2) If $a \equiv b$ (mod $n$), then $b \equiv a$ (mod $n$).

(C3) If $a \equiv b$ (mod $n$) and $b \equiv c$ (mod $n$), then $a \equiv c$ (mod $n$).

Moreover, congruence gets along very well with addition, subtraction, and multiplication:

(C4) If $a \equiv b$ (mod $n$) and $c \equiv d$ (mod $n$), then $a + c \equiv b + d$ (mod $n$), $a - c \equiv b - d$ (mod $n$), and $ac \equiv bd$ (mod $n$).

Cancellation needs an extra condition:

(C5) If $am \equiv bm$ (mod $n$) and $(m, n) = 1$, then $a \equiv b$ (mod $n$).

It is simpler to discuss cancellation when the modulus is a prime. In that case, (C5) becomes:

(C6) Suppose that $p$ is a prime. If $am \equiv bm$ (mod $p$) and $m \not\equiv 0$ (mod $p$), then $a \equiv b$ (mod $p$).

This last fact is so helpful that we will try to use a prime modulus in our congruences whenever possible.

There is one more helpful fact about congruences, and this one we'll prove, using theorem 1.1.

**THEOREM 1.3**: *Suppose that $p$ is a prime that does not divide some integer $a$. Then there is an integer $\mu$ such that $a\mu \equiv 1$ (mod $p$).*

**PROOF**: Because $(a, p) = 1$, we can find integers $\mu$ and $v$ such that $a\mu + pv = 1$. Rewrite that equation as $a\mu - 1 = pv$, and we see that $p \mid (a\mu - 1)$. In other words, $a\mu \equiv 1$ (mod $p$). $\square$

## 3. Wilson's Theorem

These ideas can be applied to yield a striking result called "Wilson's Theorem."

**THEOREM 1.4**: *Suppose that p is a prime. Then* $(p-1)! \equiv -1$ (mod *p*).

Notice that we are making an assertion about rather large numbers, even when $p$ is not that large. For example, when $p = 31$, then Wilson's Theorem asserts that $30! \equiv -1$ (mod 31), which expands out to

$$265252859812191058636308480000000 \equiv -1 \quad (\text{mod } 31),$$

or equivalently that 265252859812191058636308480000001 is a multiple of 31.

**PROOF**: Let $p$ be a prime number. We want to show that $(p-1)! \equiv -1$ (mod $p$).

We begin by listing all the positive integers from 1 to $p - 1$:

$$1, 2, 3, \ldots, p - 1.$$

Their product is $(p - 1)!$. Let $x$ be one of these numbers. Is there a $y$ on the list such that $xy \equiv 1$ (mod $p$)? Yes! That's exactly what we proved in theorem 1.3. We can take $y$ to be the number on the list that is congruent to $\mu$ mod $p$, where $\mu$ is the integer given to us by theorem 1.3.

We call $y$ the *inverse* of $x$ modulo $p$. We are justified in saying *the* inverse because $y$ is unique. Why? Suppose $xy' \equiv 1$ (mod $p$) for some other $y'$ also on the list. Then $xy \equiv xy'$ (mod $p$). Multiply through by $y$ to obtain $yxy \equiv yxy'$ (mod $p$). But $yx \equiv 1$ (mod $p$), so we conclude that $y \equiv y'$ (mod $p$). Because both $y$ and $y'$ are on the list, their difference is between 0 and $p$, and cannot be divisible by $p$. So no other inverse for $x$ can exist; $y$ is the only one.

Now we group the numbers on our list in pairs, each with its inverse. The complication is that some of the numbers might be their own inverse! When can this happen? Well, $x$ is its own inverse if and only if $x^2 \equiv 1$ (mod $p$). Equivalently, $(x - 1)(x + 1) = x^2 - 1 \equiv 0$ (mod $p$). In other words, $p$ must divide $(x - 1)(x + 1)$. Because $p$ is prime, this can happen only if

$p$ divides either $x - 1$ or $x + 1$.[3] Thus, the numbers on the list that are their own inverse are exactly 1 and $p - 1$.

So reorder our list as[4]

$$1, p - 1, a_1, b_1, a_2, b_2, \ldots, a_t, b_t,$$

where $a_i b_i \equiv 1 \pmod{p}$ for each $i$. Multiplying everything together modulo $p$, we obtain their product $\equiv p - 1 \pmod{p}$. In other words, $(p - 1)! \equiv -1 \pmod{p}$. $\qquad\square$

## 4. Quadratic Residues and Nonresidues

We start with some terminology:

**DEFINITION**: Let $p$ be a prime that is not 2. If $p$ does not divide an integer $a$ and $a \equiv b^2 \pmod{p}$ for some $b$, then $a$ is a *quadratic residue* modulo $p$. If $p$ does not divide $a$ and $a \not\equiv b^2 \pmod{p}$ for any integer $b$, then $a$ is a *quadratic nonresidue* modulo $p$.

Typically, this terminology is shortened to "residue" and "nonresidue," with the word "quadratic" and the modulus implicitly understood. In fact, for the remainder of this section, we will often omit "$(\bmod\, p)$" from our congruences to save some space.

After choosing some prime $p$, making a list of residues modulo $p$ can be done by squaring the integers from 1 to $p - 1$. But the task is actually only half as long, because $k^2 \equiv (p - k)^2 \pmod{p}$, so we only need to square the integers from 1 to $(p - 1)/2$. For example, the residues modulo 31 are

$$1, 4, 9, 16, 25, 5, 18, 2, 19, 7, 28, 20, 14, 10, 8.$$

We computed this list by squaring the integers from 1 to 15 and then dividing each by 31 and computing the remainder.

---

[3] We used Unique Factorization here: If $p$ shows up in the prime factorization of $(x - 1)(x + 1)$, it must show up in the factorization of $x - 1$ or $x + 1$ (or both). Alternatively, if $(x - 1)(x + 1) \equiv 0 \pmod{p}$, and we suppose that $x - 1 \not\equiv 0 \pmod{p}$, then by (C6), $x + 1 \equiv 0 \pmod{p}$.

[4] If $p = 2$, then our list contains the single element 1, because $1 = p - 1$.

The nonresidues are the numbers between 1 and 31 that are not on the list.

There are 15 residues modulo 31, and in general there are $(p-1)/2$ residues modulo $p$. In case you're worried, we should show you why our list can't have any duplicates: If $a_1^2 \equiv a_2^2 \pmod{p}$, then $p$ divides $a_1^2 - a_2^2$, so $p$ divides the product $(a_1 - a_2)(a_1 + a_2)$. Unique prime factorization now tells us that $p$ divides $a_1 - a_2$ or $p$ divides $a_1 + a_2$, and so either $a_1 \equiv a_2 \pmod{p}$ or $a_1 \equiv -a_2 \pmod{p}$. The second possibility is ruled out if we only square numbers from 1 to $(p-1)/2$.

We can easily see that if we multiply two residues, we get another residue: If $a_1$ and $a_2$ are residues, then $a_1 \equiv b_1^2$ and $a_2 \equiv b_2^2$, and then $a_1 a_2 \equiv (b_1 b_2)^2$. This assertion is true about ordinary squares of ordinary integers as well: $(3^2)(5^2) = 15^2$.

We can also see that if we multiply a residue and a nonresidue, we must get a nonresidue. Why? Let $a$ be a residue such that $a \equiv b^2$, and let $c$ be a nonresidue. Suppose that $ac$ is a residue. Then $ac \equiv d^2$. By assumption, $a$ is not a multiple of $p$, and therefore $b$ is also not a multiple of $p$. Use theorem 1.3 to find $\mu$ such that $\mu b \equiv 1$. Squaring tells us that $\mu^2 a \equiv 1$. Take our congruence $ac \equiv d^2$, and multiply through by $\mu^2$. We get $\mu^2 ac \equiv \mu^2 d^2$. But we just pointed out that $\mu^2 a \equiv 1$, so we get $c \equiv \mu^2 d^2$. That tells us that $c$ is a residue, which contradicts our assumption that $c$ is a nonresidue. By the way, this assertion is also true about squares of ordinary integers: square $\times$ nonsquare = nonsquare.

Now we have residue $\times$ residue = residue and residue $\times$ nonresidue = nonresidue. There's one more case to consider, and it's the one that may contradict our intuition: nonresidue $\times$ nonresidue = residue. Why is this true? The key fact is that half of the integers between 1 and $p-1$ are residues and half are nonresidues.

Suppose that $c$ is a particular nonresidue. Multiply $c$ by each integer from 1 to $p-1$; we have to get $p-1$ different answers, because $(c,p) = 1$. So we have to get the numbers from 1 to $p-1$ in a different order. Each time that we multiply $c$ by a residue, the result has to be a nonresidue. That takes care of half of the integers from 1 to $p-1$. So each time we multiply $c$ by a nonresidue, we have only $(p-1)/2$ possibilities remaining, and all of them are residues!

We can check an example of this. Notice that our list of residues modulo 31 does not include 23 or 12. We therefore know that $23 \cdot 12$ must be a residue. And sure enough, $23 \cdot 12 \equiv 28 \pmod{31}$, and 28 is on the list of residues.

## 5. The Legendre Symbol

If $a$ is an integer not divisible by $p$, we define $\left(\frac{a}{p}\right)$ to be 1 if $a$ is a quadratic residue modulo $p$ and $-1$ if it is not. We call $\left(\frac{a}{p}\right)$ the *quadratic residue symbol* or the *Legendre symbol*. We studied it in Ash and Gross (2006, chapter 7), and you can learn about it in almost any textbook on elementary number theory. We just proved that the quadratic residue symbol has a very beautiful multiplicative property:

**THEOREM 1.5**: *If $a$ and $b$ are two numbers not divisible by $p$, then $\left(\frac{a}{p}\right)\left(\frac{b}{p}\right) = \left(\frac{ab}{p}\right)$.*

In other words, the product of two quadratic residues or two quadratic nonresidues is a quadratic residue, while the product of a quadratic residue and a quadratic nonresidue is a quadratic nonresidue, which is exactly what we worked out.

It turns out that it is easy to compute $\left(\frac{-1}{p}\right)$. This is closely connected to the question of how $p$ factors in the ring of *Gaussian integers* $\mathbf{Z}[i] = \{a + ib \mid a, b \text{ integers}\}$, but we won't discuss that idea further in this book. Here's the answer:

**THEOREM 1.6**:

$$\left(\frac{-1}{p}\right) = \begin{cases} 1 & \text{if } p \equiv 1 \pmod{4} \\ -1 & \text{if } p \equiv 3 \pmod{4}. \end{cases}$$

**PROOF**: We will prove two things:

- If $\left(\frac{-1}{p}\right) = 1$, then $p \equiv 1$ (mod 4).
- If $p \equiv 1$ (mod 4), then $\left(\frac{-1}{p}\right) = 1$.

A moment's thought shows that these two things are equivalent to the theorem as stated. In both cases, the proof of the assertion depends on the fact that $p \equiv 1$ (mod 4) is the same as the statement that the integer $(p-1)/2$ is even.[5]

Suppose first that $\left(\frac{-1}{p}\right) = 1$. In other words, we know that $-1$ is a residue. Because residue × residue = residue, we know that if we take the set of residues and multiply each residue by $-1$, we get back the set of residues (in some other order).

Let's start with the residues between 1 and $(p-1)/2$. When we multiply those residues by $-1$, the result is always between $(p+1)/2$ and $p-1$, and we know that the result also has to be a residue. Conversely, if we take the residues between $(p+1)/2$ and $p-1$ and multiply by $-1$, we have to get residues between 1 and $(p-1)/2$.

This argument shows that the number of residues is *even*: Half of them lie between 1 and $(p-1)/2$, and the other half lie between $(p+1)/2$ and $p-1$. But we also know that half of the numbers between 1 and $p-1$ are residues and half are nonresidues, so we know that the number of residues is $(p-1)/2$. We have therefore shown that $(p-1)/2$ is even, which shows that $p \equiv 1$ (mod 4).

Suppose instead that $p \equiv 1$ (mod 4). We want to show that $-1$ is a residue. To show that a number $a$ is a residue, it is enough to find $b \not\equiv 0$ such that $a \equiv b^2$ (mod $p$), so we need to find some integer $b$ such that $b^2 \equiv -1$ (mod $p$).

---

[5] Why? If $p \equiv 1$ (mod 4), then $p = 4k + 1$, and so $(p-1)/2 = 2k$, which is even. Conversely, if $(p-1)/2$ is even, then $(p-1)/2 = 2k$ and $p = 4k + 1$.

We know from Wilson's Theorem (theorem 1.4) that
$(p-1)! \equiv -1 \pmod{p}$. We will now show that $(p-1)! \equiv \left[\left(\frac{p-1}{2}\right)!\right]^2$
(mod $p$). Those parentheses are so complicated that a numerical
example is called for: If $p = 29$, then we are asserting that
$28! \equiv (14!)^2 \pmod{29}$. That's the same as asserting that
$304888344611713860501504000000 \equiv (87178291200)^2 \pmod{29}$,
which can be verified by computer or a very expensive pocket
calculator.

Take $(p-1)!$, and write it out: $(p-1)! = 1 \cdot 2 \ldots (p-2)$
$(p-1)$. Let's regroup, so that we multiply the first term by the
last term, the second term by the penultimate term, and so on:

$$(p-1)! = [1 \cdot (p-1)][2 \cdot (p-2)][3 \cdot (p-3)] \ldots [(p-1)/2 \cdot (p+1)/2].$$

Notice that the second term in each pair is congruent to $-1$
times the first term: $p - 1 \equiv -1 \cdot 1 \pmod{p}, p - 2 \equiv -1 \cdot 2$
(mod $p$), and so on. So we can rewrite:

$$(p-1) \equiv [1 \cdot 1](-1)[2 \cdot 2](-1)[3 \cdot 3](-1) \ldots$$
$$\times [(p-1)/2 \cdot (p-1)/2](-1) \pmod{p}.$$

How many factors of $-1$ are present? The answer is $(p-1)/2$,
which is an even number under our hypothesis that $p \equiv 1$
(mod 4), so all of those factors of $-1$ multiply together to give 1
and we can omit them:

$$(p-1)! \equiv [1 \cdot 1][2 \cdot 2][3 \cdot 3] \ldots [(p-1)/2 \cdot (p-1)/2] \pmod{p}.$$

Now, we can rewrite one more time:

$$(p-1)! \equiv \left[\left(\frac{p-1}{2}\right)!\right]^2 \pmod{p}.$$

So we not only know that $-1$ is a residue but we have even
produced a number that solves $b^2 \equiv -1$. To return to our
numerical example, with $p = 29$, we have just shown that
$(14!)^2 \equiv -1 \pmod{29}$. Again, this assertion can be verified by
computer.                                                                      □

# Chapter 2

# SUMS OF TWO SQUARES

## 1. The Answer

You can amaze someone who enjoys playing around with numbers by having her choose a prime number and then telling her instantly whether it can be expressed as the sum of two squares. (In section 2 of the Introduction, we have precisely defined "the sum of two squares.") For example, suppose she chooses 97. You say immediately that 97 is the sum of two squares. Then she can experiment by adding pairs of numbers on the list

$$0, 1, 4, 9, 16, 25, 36, 49, 64, 81$$

(allowing a pair to contain the same square twice), and she will find that indeed 97 equals 16 plus 81. If she chooses 79, you say immediately that 79 is not the sum of two squares, and experiment shows that indeed you are correct. How do you do this trick?

**THEOREM 2.1**: *An odd prime number is the sum of two squares if and only if it leaves a remainder of 1 when divided by 4.*

We discuss the proof of this theorem later in the chapter. But even at this point, you can now do the trick yourself. It is as easy to answer the question for very large prime numbers as for small ones, because the remainder when a number in base 10 is divided by 4 depends only on the last two digits of the number. (PROOF: Take the number $n = ab \ldots stu$, where $a, b, \ldots, s, t, u$ are decimal digits. Then $n = 100(ab \ldots s) + tu$. Therefore, $n = 4(25)(ab \ldots s) + tu$, and so the remainder when you divide $n$ by 4 is the same as the remainder when you divide $tu$ by 4.) However, unless you have a very rare ability, you can't tell at a glance whether a large number is prime or not.

Now we have computers or the Internet to help us. For example, a few seconds of work on the Internet told us that 16561 is prime. Because 61 leaves a remainder of 1 when divided by 4, we know that 16561 does also. Therefore, by our theorem, 16561 is the sum of two squares. If we want to find two squares that add to 16561, we'd have to run the computer program we imagined in the Introduction. When we run the program, we find that 16561 equals $100^2 + 81^2$.

By the way, if our computer program goes through all possibilities, it will tell us that this is the *only* way to write 16561 as a sum of two squares (except for the trivial other way: $16561 = 81^2 + 100^2$). Fermat already knew that if a prime was the sum of two squares, then it was so in a unique way (except for the trivial exchange). He also knew theorem 2.1. Perhaps he had proofs of these things, but the first published proofs are due to Euler.

What can we say about a nonprime number $n$? When is $n$ a sum of two squares? This question gives us a good example of how to reduce a problem to a simpler or more basic problem. Just staring at the question won't help. But we can get started by using the following formula:

$$(x^2 + y^2)(z^2 + w^2) = (xz - yw)^2 + (xw + yz)^2. \qquad (2.2)$$

This formula holds for whatever numbers you choose to substitute for the variables $x$, $y$, $z$, and $w$. You can check it by multiplying out both sides and doing some algebra.

**DIGRESSION**: Where did this formula come from? After we write down the formula, you can easily multiply out both sides and see that the equality holds. But how could someone discover such a formula in the first place? One way might be just playing around with algebra. Another way might be experimental: Multiply lots of pairs of sums of two squares and notice how the product is also a sum of two squares and in what way.

A third way is by using complex numbers. They are not a prerequisite for reading this part of the book, but if you

happen to know about them, you can derive the formula.
Take the formula for the product of two complex numbers:
$(x + iy)(z + iw) = (xz - yw) + i(xw + yz)$. Then take the norm of
both sides and use the fact that $|ab| = |a||b|$ for any two complex
numbers $a$ and $b$.

Notice that the field of complex numbers is "two-dimensional
over the real numbers." That means that each complex number
is made out of two independent real numbers, so that $2 + 3i$ is
made out of 2 and 3, for instance. The "two" dimensions is the
same "two" as the "two" in the problem we are studying: sums of
*two* squares. It is not a coincidence that the wonderfully
powerful complex numbers can be used to study sums of *two*
squares. Compare what happens in the next chapter, when we
try to study sums of three squares.

Equation (2.2) tells us that if $A$ and $B$ are each the sum of two
squares, then so is their product $AB$. It even tells us how to find
two squares that add up to $AB$ if we know such data for $A$ and
$B$ separately. For example, we saw that $97 = 4^2 + 9^2$. It is easy to
see that $101 = 1^2 + 10^2$. So we can conclude that $97 \cdot 101 = (4 \cdot 1 -
9 \cdot 10)^2 + (4 \cdot 10 + 9 \cdot 1)^2$ or $9797 = 86^2 + 49^2$. CHECK: $86^2 = 7396$,
$49^2 = 2401$, and $7396 + 2401 = 9797$. We have parlayed some fairly
obvious small number facts into a more surprising large number
fact.

Our theorem told us that an odd prime was a sum of two squares
if and only if it left a remainder of 1 when divided by 4. Let's
temporarily name two kinds of odd primes. If a prime leaves a
remainder of 1 when divided by 4, let's call it a "Type I prime."
If a prime leaves a remainder of 3 when divided by 4, let's call it
a "Type III prime." (This terminology is not standard, and we will
not use it again after this chapter.) Note also that 2 is the sum of
two squares: $1 + 1$. Also, any square number $s$ by itself is the sum
of two squares: $0 + s$.

Suppose now that we are given the positive integer $n$. Factor it
into primes,

$$n = 2^a 3^b 5^c \cdots p^t,$$

where the exponents $a$, $b$, $c, \ldots, t$ are nonnegative integers, possibly 0.

Every factor of 2 is the sum of two squares. Every Type I prime in the factorization of $n$ is also the sum of two squares. Suppose that every Type III prime in the factorization of $n$ has an even exponent. Then $n$ is the sum of two squares.

Why is this claim true? It's easiest to explain by example. Suppose

$$n = 2 \cdot 3^4 \cdot 5^3.$$

We can write this as

$$n = 2 \cdot 9 \cdot 9 \cdot 5 \cdot 5 \cdot 5.$$

Now 2 and 9 are each the sum of two squares ($2 = 1^2 + 1^2$ and $9 = 3^2 + 0^2$), so by our formula, the same is true of their product $2 \cdot 9$. Next, since $2 \cdot 9$ and 9 are each the sum of two squares, again by our formula, the same is true of the product $2 \cdot 9 \cdot 9$.

We continue merrily on our way. Because 5 is the sum of two squares, we conclude that $2 \cdot 9 \cdot 9 \cdot 5$ is also the sum of two squares. Then we have the same conclusion about $2 \cdot 9 \cdot 9 \cdot 5 \cdot 5$ and finally for $2 \cdot 9 \cdot 9 \cdot 5 \cdot 5 \cdot 5$. You see, we have to go one step at a time. We could automate this using "mathematical induction"—see chapter 5 for an illustration of how to use induction in proofs. (By the way, you may want to follow up this example. The number $n$ is 20250. Run your program to find a way to write 20250 as a sum of two squares. Or, more painstakingly, use our formula repeatedly with inputs $2 = 1^2 + 1^2$, $9 = 0^2 + 3^2$, and $5 = 1^2 + 2^2$ to find how to write 20250 as a sum of two squares.)

So we can now state a theorem: A positive integer is a sum of two squares if every Type III prime occurring in its prime factorization occurs with an even exponent. Note that a Type III prime (in fact, any integer) raised to an even exponent is a square: $a^{2k} = (a^k)^2$.

This is only provisional, because we would like to have an "if and only if" statement. Could the statement actually be "if and only if"? On the other hand, in theorem 2.1, the key property was the remainder a number left when divided by 4. We might think this is true in general—that if a number leaves the remainder 1 when divided by 4, then it will be a sum of two squares, even if it has Type III primes with odd exponents.

We can try an experiment. The number $21 = 3 \cdot 7$ is the product of two Type III primes, each with an odd exponent. But 21 does leave a remainder of 1 when divided by 4. Perhaps 21 is the sum of two squares? Alas, trial and error shows that 21 is *not* the sum of two squares, even though it leaves the remainder 1 when divided by 4. After doing some more experiments, we would be led to guess that "if and only if" *is* appropriate. In fact, we can state:

**THEOREM 2.3**: *A positive integer n is a sum of two squares if and only if every Type III prime that occurs in the prime factorization of n occurs with an even exponent.*

We will discuss the proofs of theorems 2.1 and 2.3 in the rest of this chapter.

## 2. The Proof Is Not in the Pudding

The "pudding" refers to practical experience. In our case, this means writing computer programs or doing calculations by hand, making lists of numbers, their prime factorizations, and whether they are sums of two squares or not. We could do billions of experiments and become convinced of the truth of theorems 2.1 and 2.3. Mathematicians are interested in such experiments and often perform them. But mathematicians do not accept such experimental results as definitive.

There are famous examples in number theory of statements that are true for billions of examples but turn out to be provably false. One of the most famous concerns a "race" between Type I and Type III primes. Take any number $n$. Count the number of Type I primes less than $n$ and the number of Type III primes less than $n$. There always seem to be more Type III primes, at least if $n$ is less than a billion. But eventually the Type I primes will take the lead at least briefly. For a beautiful survey of this and many other examples of statements that are true for "small" numbers but not always true, see Guy (1988).

That's why we need to find proofs of theorems 2.1 and 2.3. The details become rather complicated. We will discuss the most interesting parts of the argument first and leave the rest of the details to section 4.

First, let's prove the "only if" parts of theorems 2.1 and 2.3. In other words, we want to prove the following:

Let $n$ be a positive integer such that there is at least one Type III prime appearing in its prime factorization with an odd exponent. Then $n$ is not the sum of two squares.

For example, if $n$ itself is an odd prime, the hypothesis says that $n$ is a Type III prime, and the statement would boil down to the "only if" part of theorem 2.1.

Rather than give a complete proof of the entire assertion, we will demonstrate the main idea by proving that if

(1) $n$ is a sum of two squares
and
(2) if $q$ is a Type III prime that divides $n$,
then
(3) the exponent of $q$ in the prime factorization of $n$ must be at least 2.

In other words, $q^2$ divides $n$.

The ideas used in this proof can then be used further to show that the exponent of $q$ in the prime factorization of $n$ is even by replacing $n$ with $n/q^2$ and using mathematical induction.

Assume (1) and (2). Then there exist integers $a$ and $b$ with the property that $n = a^2 + b^2 \equiv 0 \pmod{q}$. If $q$ divides both $a$ and $b$, then $q^2$ divides both $a^2$ and $b^2$ and hence $q^2$ divides their sum $n$. In that case, (3) is true. So we will show that $q$ divides both $a$ and $b$. Equivalently, we will assume that $q$ does not divide one of $a$ or $b$ and derive a contradiction. (This is proof by "reductio ad absurdum.")

It doesn't matter which of the two numbers we call $a$ and which one $b$, so we may as well assume that $q$ does not divide $b$. This implies that $q$ and $b$ are relatively prime. An application of theorem 1.3 gives us an integer $\lambda$ such that $\lambda b \equiv 1 \pmod{q}$. Multiplying this congruence by itself, we derive the important provisional fact that

$$\lambda^2 b^2 \equiv 1 \pmod{q}. \tag{2.4}$$

Multiply the congruence $a^2 + b^2 \equiv 0 \pmod{q}$ through by $\lambda^2$ and use (2.4). We deduce another fact that under the same assumption that $q \nmid b$, then

$$\lambda^2 a^2 + 1 \equiv 0 \pmod{q}.$$

Now $\lambda a$ is some integer; for simplicity, let $\lambda a = c$.

Summarizing our work so far, we have seen that if we assume (1) and (2), then there exists an integer $c$ with the property that $c^2 \equiv -1 \pmod{q}$. But now theorem 1.6 tells us that $q \equiv 1 \pmod 4$; that is, *q is a Type I prime, not a Type III prime.* That's a contradiction to hypothesis (2) and completes our proof.

This argument was rather intricate. The contradiction just sort of sneaks up on you. The more conceptual way of doing this proof involves a little bit of group theory, and we don't want to assume any knowledge of group theory until Part III.

## 3. The "If" Parts of Theorems 2.1 and 2.3

We already saw in section 1 of this chapter that the "if" part of theorem 2.1 implies the "if" part of theorem 2.3. So we only need to worry about writing a Type I prime as the sum of two squares.

Let's suppose $p$ is a Type I prime. How can we prove that $p$ is the sum of two squares? Taking a cue from the preceding work, we notice that if $a^2 + b^2 = p$, then $a^2 + b^2 \equiv 0 \pmod{p}$, and arguing with $\lambda$ as before, we derive that $-1$ is a square modulo $p$. Our first order of business is to reverse this. We know from theorem 1.6 that if $p$ is a positive prime that leaves a remainder of 1 when divided by 4, then there exists an integer $c$ with the property that $c^2 \equiv -1 \pmod{p}$.

Of course, we're not done yet. All we have done so far is to show that if $p$ is a Type I prime, then there exists an integer $c$ such that $c^2 \equiv -1 \pmod{p}$. Let's sketch the rest of the proof from here on.

First of all, the congruence means that there is some positive integer $d$ such that $c^2 + 1^2 = pd$. So at least some multiple of $p$ is a sum of two squares. Euler argued by "descent": Given that $dp$ is a sum of two squares, and $d > 1$, show algebraically that there is a strictly smaller positive $d'$ such that $d'p$ is also a sum of two squares. Keep working until you reduce $d$ down to 1. We will explain this argument in the next section.

## 4. The Details

The proofs in this section are standard. We follow the exposition in Davenport (2008), which we recommend highly as a first textbook in elementary number theory.

**GIVEN**: There exist positive integers $c$ and $d$ such that $c^2 + 1^2 = pd$.

**TO PROVE**: There exist positive integers $a$ and $b$ such that $a^2 + b^2 = p$.

**PROOF**: First, note that if we set $C = c - kp$ for any integer $k$, then $C \equiv c \pmod{p}$, and so

$$C^2 + 1 \equiv c^2 + 1 \equiv 0 \pmod{p}.$$

By choosing $k$ properly, we may achieve $|C| < p/2$. (This is called "dividing $c$ by $p$ and finding the remainder $C$," except we allow $C$ to be negative.) If $C$ is negative, we may replace it by its absolute value, because all that we care about is that $C^2 + 1 \equiv 0 \pmod{p}$. Thus, we may derive from what's given to conclude that:

There exist positive integers $C$ and $d$ such that $C < p/2$ and $C^2 + 1^2 = pd$.

In this case, we see that $d$ cannot be too big. In fact, $pd = C^2 + 1 < (p/2)^2 + 1 = p^2/4 + 1$, which implies that $d < p$. Notice that if $d = 1$, we're finished: We've written $p$ as a sum of two squares. So we suppose that $d > 1$.

We start our descent with the following statement:

There exist positive integers $x, y$, and $d$ such that

$$1 < d < p \text{ and } x^2 + y^2 = pd. \tag{2.5}$$

We will prove that assuming (2.5), there exist positive integers $X, Y$, and $D$ such that $D < d$ and $X^2 + Y^2 = pD$. When we succeed in doing this, it will follow that we can keep repeating this implication until the new $D$ is forced down to becoming $D = 1$. Thus we will have proven the statement "TO PROVE."

The trick here is to use congruences modulo $d$. From (2.5), we know that $x^2 + y^2 \equiv 0 \pmod{d}$. It is funny how $d$ and $p$ sort of trade places in our reasoning here. This congruence will still hold if we replace $x$ by any $u$ congruent to $x \pmod{d}$ and $y$ by any $v$ congruent to $y \pmod{d}$. We may choose $u$ and $v$ such that $0 \le u, v \le d/2$. (Compare what we did with $c$ and $C$ earlier.) Note that $u$ and $v$ cannot both be 0. That would imply that $d$ divides both $x$ and $y$, which would imply that $d^2$ divides $x^2 + y^2 = pd$, which is impossible because $p$ is prime and $d < p$.

We have $u^2 + v^2 = ed$ for some positive integer $e$ and $u, v \le d/2$. How big can $e$ be? Well, $ed = u^2 + v^2 \le (d/2)^2 + (d/2)^2 = d^2/2$, so $e \le d/2$. In particular, $e < d$. It will turn out that $e$ will be our new $D$, so we will be in good shape.

Now we use (2.2). We know that $x^2 + y^2 = pd$ and that $u^2 + v^2 = ed$. Multiplying, we get

$$A^2 + B^2 = (x^2 + y^2)(u^2 + v^2) = ped^2,$$

where $A = xu + yv$ and $B = xv - yu$. But look at $A$ and $B$ $\pmod{d}$. We have $A = xu + yv \equiv xx + yy = x^2 + y^2 = pd \equiv 0$ $\pmod{d}$ and $B = xv - yu \equiv xy - yx = 0 \pmod{d}$. This is great because that means we can divide the last displayed equation through by $d^2$, obtaining

$$(A/d)^2 + (B/d)^2 = pe.$$

We've accomplished our goal, because we can take $X = \frac{A}{d}$, $Y = \frac{B}{d}$, and $D = e$. $\qquad\square$

This is a very neat proof. You can see that it cleverly takes into account the identity for the product $(x^2 + y^2)(z^2 + w^2)$. There are exactly *two* terms in this product, each is the sum of exactly *two* squares, and a square is a number raised to the power *two*. This repetition of *two* makes everything work. For instance, if you try to do this with sums of three squares or sums of two cubes, it is not going to work, at least not in this easy way. We touch lightly on three squares in the next chapter. Techniques for dealing with two cubes are very different and beyond the scope of this book.

*Chapter 3*

# SUMS OF THREE AND FOUR SQUARES

## 1. Three Squares

The question of which numbers are the sums of three squares is very much more difficult to answer:

**THEOREM 3.1**: *A positive integer n is the sum of three squares if and only if n is not equal to a power of* 4 *times a number of the form* $8k + 7$.

It's easy to see that numbers of the form $8k + 7$ cannot be written as sums of three squares. If $n = 8k + 7$, then $n \equiv 7 \pmod 8$. But a little bit of squaring will show you that every square integer is congruent to either 0, 1, or 4 (mod 8). So three squares could not add up to 7 (mod 8) and hence could not add up to $n$. Proving the whole theorem is quite difficult, and the proof is not given in such elementary textbooks as Davenport (2008) and Hardy and Wright (2008).

Why are three squares so much more difficult than two or four squares? One answer is that the product of a pair of sums of two squares is itself a sum of two squares, as we saw in formula (2.2). There is another such formula for a pair of sums of four squares, which we will give later in this chapter. There can be no such formula for sums of three squares. Indeed, $3 = 1^2 + 1^2 + 1^2$ and $5 = 0^2 + 1^2 + 2^2$ are each sums of three squares, but their product $15 = 8 \cdot 1 + 7$ is not.

Still, you might ask, *why* is there no such formula? In fact, a difficult theorem says that such formulas can only exist for sums of two, four, and eight integral squares.

Later, when we discuss "in how many different ways is $n$ a sum of $b$ squares" for general $b$, again the differences between the case when $b$ is even and when $b$ is odd will be very striking. Again, the case of odd $b$ will be so difficult that we will merely mention it without giving any further treatment. We will be able to give an analytic reason for the difference between even and odd numbers of squares, but there doesn't seem to be any elementary explanation.

## 2. Interlude

Before going on to four squares, we want to make a few remarks. First, notice that our proof that $8k + 7$ was not the sum of three squares won't work with four squares. We can get 7 (mod 8) by adding $1 + 1 + 1 + 4$. So that is a hint that every number might just be a sum of four squares. In fact, it's true:

**THEOREM 3.2**: *Every positive number is the sum of four squares.*

Second, we state the formula:

$$(a^2 + b^2 + c^2 + d^2)(A^2 + B^2 + C^2 + D^2)$$
$$= (aA + bB + cC + dD)^2 + (aB - bA + cD - dC)^2$$
$$+ (aC - cA + dB - bD)^2 + (aD - dA + bC - cB)^2. \quad (3.3)$$

You can verify (3.3) by brute force: Just multiply out both sides. If you know about quaternions, then you can see that this formula expresses the fact that the norm of a product of quaternions is the product of their norms. It follows from (3.3) that to prove the theorem, all we need to show is that every prime is a sum of four squares.

# 3. Four Squares

Now we discuss the proof of the four squares theorem, which was found originally by Lagrange. Because of equation (3.3), it will be enough if we can show that any prime $p$ is the sum of four squares. So let $p$ be a prime.

If $p$ is 2, we're all set because $2 = 0^2 + 0^2 + 1^2 + 1^2$. Now let's assume that $p$ is odd. If $p$ leaves a remainder of 1 when divided by 4, we're finished because we know $p$ is the sum of two squares, and then we can just add on $0^2 + 0^2$.

We might now assume that $p$ leaves a remainder of 3 when divided by 4 and try to be clever. We could try to subtract from $p$ a smaller prime $q$ with the properties that

(1) $q$ leaves a remainder of 1 when divided by 4, so it is the sum of two squares; and

(2) $p - q$ is also a sum of two squares (which from chapter 2 means that every Type III divisor of $p - q$ appears in the prime factorization of $p - q$ with an even exponent).

More generally, we could take for $q$ any number that is itself a sum of two squares.

Well, that's being too clever. It doesn't seem to work very well. Subtracting one number from another usually doesn't let you figure out the prime factors of the difference in any easy way.

Let's try again. We still assume $p$ is an odd prime that leaves a remainder of 3 upon division by 4. To get an idea of what to do, suppose $p$ is a sum of four squares. Say $p = a^2 + b^2 + c^2 + d^2$. Not all of $a, b, c,$ and $d$ can be 0, obviously. (For once, something "obvious" is obvious.) But they are all less than $p$, equally obviously. So we may suppose that $a$ is strictly between 0 and $p$. Then $a$ has an inverse $w$ mod $p$. We can multiply the congruence $0 \equiv a^2 + b^2 + c^2 + d^2$ (mod $p$) through by $w^2$, as we did in the previous chapter, and subtract $-1$ from both sides. We get the fact that if $p$ is the sum of four squares, then $-1$ is the sum of three squares, modulo $p$.

Our experience with *two* squares told us that showing that $-1$ was the sum of *one* square mod $p$ was a good way to start. Here, too, because $-1$ being a sum of three squares  (mod $p$) is a necessary condition for $p$ to be a sum of four squares, we may try to start by proving that and then using a descent argument to finish the proof.

**LEMMA 3.4**: *If $p$ is an odd prime that leaves a remainder of 3 upon division by 4, then $-1$ is the sum of three squares* (mod $p$).

**PROOF**: In fact, we can show that $-1$ is the sum of *two* squares (mod $p$). We know that 1 is a square (i.e., a quadratic residue) and $p - 1$ (i.e., $-1$) is a nonresidue (mod $p$). Therefore, if we start at 1 and count upwards there must be some point at which some integer $n$ is a quadratic residue and $n + 1$ is a nonresidue. We can therefore write $n \equiv x^2$ (mod $p$), and we know that $x^2 + 1$ is a nonresidue.

We saw in chapter 1, section 4, that nonresidue × nonresidue = residue. Therefore, $(-1)(x^2 + 1)$ is a residue. Write $-(x^2 + 1) \equiv y^2$ (mod $p$), and then we can see that $-1 \equiv x^2 + y^2$ (mod $p$).     □

Using Lemma 3.4, we can write $-1 \equiv a^2 + b^2 + c^2$ (mod $p$), and consequently $0 \equiv a^2 + b^2 + c^2 + d^2$ (mod $p$), for some integers $a$, $b$, $c$, and $d$ (where in fact $d = 1$). Then there exists $m > 0$ such that $a^2 + b^2 + c^2 + d^2 = pm$. By taking $a$, $b$, $c$, and $d$ between $-p/2$ and $p/2$, we can ensure that $m$ is less than $p$.

We can then construct an argument just like the descent we made in chapter 2 but algebraically a bit more complicated. That is, from the equality $a^2 + b^2 + c^2 + d^2 = pm$, we could now derive a new equality giving $pM$ as the sum of four squares but with a positive $M$ strictly less than $m$. The key idea is similar to what we did in chapter 2, section 4, now using (3.3), and again dividing out by $m^2$.

## 4. Sums of More Than Four Squares

It might seem dumb to consider more than four squares. If $n$ is the sum of four squares, which it is for any positive integer $n$, then it is the sum of 24 squares also—just add on a bunch of $0^2$'s. But we can still ask a very interesting question here, which then will also apply to sums of two, three, or four squares. Namely:

Given the positive integers $k$ and $n$, *in how many different ways* can $n$ be written as a sum of $k$ squares?

In chapter 10, section 2, we will explain exactly how we count the different ways and how this problem leads us onward into the theory of generating functions and modular forms.

Chapter 4

# SUMS OF HIGHER POWERS: WARING'S PROBLEM

## 1. $g(k)$ and $G(k)$

We have just seen that every positive integer is the sum of four squares, where we count 0 as a square. Many generalizations immediately present themselves. Of these, one of the most interesting derives from an assertion of Edward Waring in 1770 that every number is the sum of 4 squares, of 9 cubes, of 19 biquadrates (fourth powers), and so on. The most interesting part of Waring's statement is "and so on": It implies that after picking a positive integer $k$, one can find some other integer $N$ such that every positive integer is a sum of $N$ nonnegative $k$th powers. It seems extremely unlikely that Waring could have had a proof in mind, given that the first proof was published by Hilbert in 1909.

It is customary to use the notation $g(k)$ to represent the smallest integer $N$ such that every positive integer is the sum of $N$ nonnegative $k$th powers. (To avoid unnecessary repetition, for the rest of this chapter, when we refer to $k$th powers, we mean nonnegative $k$th powers.) In this language, the equation $g(2) = 4$ means simultaneously that:

- Every integer is the sum of four squares.
- There are integers that cannot be written as the sum of three squares.

We showed in the previous chapter how to prove the first of these assertions. The second assertion follows by verifying from trial and error that 7 cannot be written as the sum of three squares.

Waring's assertions are first that $g(3) \leq 9$ and $g(4) \leq 19$, and second that $g(k)$ is finite for every positive integer $k$. It is not hard to check that the number 23 cannot be written as the sum of eight cubes. On the other hand, $23 = 2 \cdot 2^3 + 7 \cdot 1^3$, so 23 is the sum of nine cubes. Similarly, $79 = 4 \cdot 2^4 + 15 \cdot 1^4$, and again trial and error shows that 79 is not the sum of 18 biquadrates. Waring of course was aware of these calculations, and so his first assertion is actually that $g(3) = 9$ and $g(4) = 19$.

Something more interesting is true of sums of cubes and higher powers that is not true about squares. Consider first the problem of cubes. We saw that 23 is the sum of nine cubes and no fewer. Some more work shows that $239 = 2 \cdot 4^3 + 4 \cdot 3^3 + 3 \cdot 1^3$, and again that 239 cannot be written as the sum of fewer than nine cubes. However, it can be proven that these are the only two numbers that cannot be written as the sum of eight cubes.

In this situation, mathematicians decided to treat these facts about 23 and 239 as numerical oddities: There are not enough cubes available to write those two numbers as sums of eight cubes. However, all other numbers can be written as a sum of eight cubes. This assertion is regarded by experts as deeper and more interesting than the assertion that all numbers are the sum of nine cubes; it is also considerably harder to prove. The notation $G(k)$ is used to stand for the smallest integer $N$ such that all sufficiently large integers can be written as the sum of $N$ nonnegative $k$th powers.

In the case of squares, we know not just that $g(2) = 4$ but also that $G(2) = 4$. Why? We noted that 7 cannot be written as the sum of three squares, but it is easy to prove more. This was already stated in a different guise in chapter 3, but we repeat the very short argument:

**THEOREM 4.1**: *If $n \equiv 7$ (mod 8), then $n$ is not the sum of three squares.*

**PROOF**: Squaring the numbers $0, 1, \ldots, 7$ shows that if $a$ is any integer, then $a^2 \equiv 0$, 1, or 4 (mod 8). Suppose that $n = a^2 + b^2 + c^2$. Then $n \equiv a^2 + b^2 + c^2$ (mod 8), and trial and error shows that $n \not\equiv 7$ (mod 8). $\qquad\square$

Our earlier assertions about sums of cubes state that $G(3) \leq 8$. In fact, a bit of computer programming shows that of all positive integers less than $10^6$, very few must be written as the sum of eight nonzero cubes, and the largest of these is 454. Stated another way, all numbers between 455 and $10^6$ can be written as the sum of no more than seven cubes. It is tempting to guess that $G(3) \leq 7$. This inequality can be proven, and an elementary proof of a partial result is in Boklan and Elkies (2009). However, the precise value of $G(3)$ is not known.

It is, however, easy to see that $G(3) \geq 4$, reasoning in the same way as in theorem 4.1.

**THEOREM 4.2**: *If $n \equiv \pm 4$* (mod 9), *then $n$ is not the sum of three cubes.*

**PROOF**: If $a$ is any integer, cubing the numbers $0, 1, \ldots, 8$ shows that $a^3 \equiv 0, 1$, or $-1$ (mod 9). Therefore, if $n = a^3 + b^3 + c^3$, then $n \not\equiv \pm 4$ (mod 9). $\qquad\square$

It is known therefore that $4 \leq G(3) \leq 7$. Computer experimentation shows that relatively few numbers smaller than $10^9$ cannot be written as the sum of six cubes, suggesting that $G(3) \leq 6$. Some experts are bold enough to conjecture based on the numerical evidence that $G(3) = 4$.

## 2. Sums of Biquadrates

Because biquadrates are squares of squares, it is possible to give an elementary proof that $g(4)$ is finite. We follow the argument in Hardy and Wright (2008).

**THEOREM 4.3**: *$g(4)$ is at most* 53.

**PROOF**: Tedious computer verification gives the algebraic identity

$$6(a^2 + b^2 + c^2 + d^2)^2 = (a+b)^4 + (a-b)^4 + (c+d)^4 + (c-d)^4$$
$$+(a+c)^4 + (a-c)^4 + (b+d)^4 + (b-d)^4$$
$$+(a+d)^4 + (a-d)^4 + (b+c)^4 + (b-c)^4.$$

Therefore, any number of the form $6(a^2 + b^2 + c^2 + d^2)^2$ can be written as the sum of 12 biquadrates. Because any integer $m$ can be written in the form $a^2 + b^2 + c^2 + d^2$, we know that any integer of the form $6m^2$ is the sum of 12 biquadrates.

Now, any integer $n$ can be written in the form $6q + r$, where $r$ is 0, 1, 2, 3, 4, or 5. The number $q$ can be written as $m_1^2 + m_2^2 + m_3^2 + m_4^2$, and therefore $6q$ is the sum of 48 biquadrates. The largest possible value for the remainder $r$ is 5. Because $5 = 1^4 + 1^4 + 1^4 + 1^4 + 1^4$, the result follows. $\square$

We can also get a result for biquadrates similar to theorem 4.1:

**THEOREM 4.4**: *If $n \equiv 15$ (mod 16), then $n$ is not the sum of* 14 *biquadrates.*

**PROOF**: Some rather unpleasant calculation shows that $a^4 \equiv 0$ or 1 (mod 16). Therefore, if $n = a_1^4 + a_2^4 + \cdots + a_{14}^4$, then $n \not\equiv 15$ (mod 16). $\square$

In other words, we see that $G(4) \geq 15$. Trial and error shows that 31 is not the sum of 15 biquadrates, and a variation of theorem 4.4 can be used to show that $16^m \cdot 31$ is not the sum of 15 biquadrates for any integer $m$. Hence, $G(4) \geq 16$.

In fact, this lower bound is the actual value, $G(4) = 16$, a result proved by Davenport in 1939.

## 3. Higher Powers

It is not hard to get a nontrivial lower bound for $g(k)$. The trick is to pick a number $n$ smaller than $3^k$, so that $n$ must be written as a sum of $a \cdot 1^k$ and $b \cdot 2^k$. To be precise, let $q$ be the greatest integer less than the quotient $3^k/2^k$. Let $n = q \cdot 2^k - 1$, so we see immediately that $n < 3^k$. Some thought shows that $n = (q - 1)2^k + (2^k - 1)1^k$, so $n$ is the sum of $(q - 1) + (2^k - 1)$ $k$th powers. We get:

**THEOREM 4.5**: $g(k) \geq 2^k + q - 2.$

The experts conjecture that in fact $g(k) = 2^k + q - 2$. For example, $g(4) = 2^4 + 5 - 2$, and $q$ in this case is 5, because $3^4/2^4 = 81/16 = 5 + \frac{1}{16}$. This equality for $g(k)$ is now known to be correct for all but finitely many values of $k$ (Mahler, 1957).

As noted earlier, the value of $G(k)$ is thought to be more significant, because it is independent of small numerical oddities. The mathematics involved is also much harder. The first results were obtained by Hardy and Littlewood, using what they called the *circle method*. The method was improved by Vinogradov, who showed in 1947 that

$$G(k) \leq k(3 \log k + 11).$$

The study of this problem remains an active research topic.

*Chapter 5*

# SIMPLE SUMS

## 1. Return to First Grade

We teach children how to add. They memorize the addition table up to $9 + 9$, or at least they used to. Then they can use place notation to add numbers of any size, at least in theory. The power of mathematics should make us pause here a second. There are things in mathematics that can be proved but never fully exemplified. For example, our students learn that addition is commutative: $x + y = y + x$ for any two integers $x$ and $y$. They check this with small examples, like $23 + 92 = 92 + 23$. But commutativity of addition is true for *any* two integers. There are integers so big that you couldn't write them down or read them in your lifetime. Take two of those—their sum in either order will be the same. (And all but finitely many numbers are this big.)

Suppose you took two fairly large numbers $a$ and $b$ that you could add together by hand, say in a year. So this year you add $a + b$ and next year you add $b + a$. We bet that you won't get the same answer. But it will be because you got tired and made a mistake in addition, not because the commutative law fails to hold for some very large numbers. Mathematics is just not an empirical science.

Back to smaller numbers. Any number written in base 10 is a sum. For example, the number 2013 stands for $2000 + 10 + 3$. This allows us to use *casting out nines* to check our addition. Casting out nines from a number (in base 10) means adding the digits together, then adding the digits of the answer together, and continuing until you have only one digit. At any stage, you can throw away a 9 or any bunch of digits that add up to 9, so at the end you will have one of the digits 0, 1, 2, 3, 4, 5, 6, 7, or 8.

To check the addition of $a + b = c$ (with all three numbers written in base 10), you cast out the nines from $a$, $b$, and $c$. You add the cast-up digits from $a$ and $b$ and you cast out nines from that sum. You should get the cast-up digit from $c$. For example,

$$2013 + 7829 = 9842.$$

CHECK: 2013 yields 6 and 7829 yields 8 from casting out nines. We add 6 and 8 to get 14, which has two digits, which give $1 + 4 = 5$. Then we check with the supposed sum 9842—yes, it yields 5 also. So there is a good chance we did the addition correctly. After all, an error would most likely just affect one digit, and this would destroy the agreement with the cast-out sum (unless the digit error were a swap of 0 and 9). Of course, if you use this to check a very long sum, with many addends, then the chance of a spurious agreement rises. It is still useful for normal everyday problems—if you don't use a calculator, or even if you do (because you might err in entering the numbers in your calculator). Casting out nines can also be used to check subtraction and multiplication.

Why does casting out nines work? A number in base 10 is a sum. For example, $abcd = 10^3 \cdot a + 10^2 \cdot b + 10 \cdot c + d$. Now $10 \equiv 1$ (mod 9) and therefore $10^k \equiv 1$ (mod 9) for any positive integer $k$. Hence $abcd \equiv a + b + c + d$ (mod 9). What we are doing when we cast out nines is simply finding the remainder of a number when divided by 9. What we are checking is that if $x + y = z$ on the nose, then it better be true that $x + y \equiv z$ (mod 9).

## 2. Adding Small Powers

Let's move on to adding consecutive powers of integers. If we start with zeroth powers, we have the formula

$$1 + 1 + \cdots + 1 = n$$

Figure 5.1. Triangular numbers

if there are $n$ terms on the left-hand side. That wasn't hard.[1] Perhaps that equation can be thought of as the definition of the integer $n$, or maybe it expresses the multiplication $1 \cdot n = n$. We will skip such philosophical matters and move on to first powers.

$$1 + 2 + 3 + \cdots + n = \frac{n(n+1)}{2}. \tag{5.1}$$

The result is called the $n$th triangular number, for reasons to be seen in figure 5.1.

How could we prove this formula? It is said that Gauss as a small child figured this out[2] as follows: Group 1 and $n$ to give $n + 1$. Group 2 and $n - 1$ to give $n + 1$. Continue. How many groupings will you have? You will have $\frac{n}{2}$ groupings, because there are $n$ numbers and each group has two partners. In total, you have $\frac{n}{2}$ times $n + 1$, and this is the claimed result. (We hope you are bothered by this argument if $n$ is odd. In that case, you need to modify it a bit, which we leave for you as an exercise.)

A more formal way to prove (5.1) is to use mathematical induction. In science, "induction" means looking at lots of occurrences of something in experience and noticing regularities. This is *not* what induction means in mathematics. In our subject, induction is a particular way to prove a statement that depends on a variable $n$, where $n$ is a positive integer. Or you can think of it as proving an infinite number of statements, each of which is indexed by an integer $n$. How does induction work?

---

[1] However, see the quotation from *Through the Looking-Glass* on page vi.

[2] For more on this story, see section 1 of chapter 6.

Mathematical induction depends on the following axiom (which we must accept as true):

If you have a set $S$ of positive integers, *and* if $S$ contains the integer 1, *and* if the statement

(∗)     If $S$ contains all the integers up to every positive integer $N - 1$, then it also contains the integer $N$

is true, then $S$ is the set of all positive integers.

So now suppose you have a statement to prove that depends on the positive integer $n$. Let's write $P_n$ for the version of the statement for the integer $n$. (The letter P here is the first letter of "proposition.") For example, say the statement we want to prove is (5.1):

$$1 + 2 + 3 + \cdots + n = \frac{n(n+1)}{2}.$$

Then $P_4$ would be the statement $1 + 2 + 3 + 4 = \frac{4(4+1)}{2} = 10$ (which is true, by the way).

Let $S$ be the set of all positive integers $m$ for which $P_m$ is true. Then, to show that $P_n$ is true for all $n$, we have to show that $P_1$ is true and

(†) if $P_k$ is true for all $k < N$ for any integer $N \geq 2$, then $P_N$ is true.

Let's try this with the problem at hand. First, we check $P_1$, which asserts that $1 = \frac{1(1+1)}{2}$. Yes, that's true.

Now, let's try the (†) step. We let $N$ be any positive integer, and we assume that $P_k$ is true for all $k < N$. We need to show that under that hypothesis (called the "inductive hypothesis"), it follows that $P_N$ is true. Now our hypothesis in particular tells us that $P_{N-1}$ is true. In other words,

$$1 + 2 + 3 + \cdots + N - 1 = \frac{(N-1)(N-1+1)}{2} = \frac{(N-1)N}{2}. \quad (5.2)$$

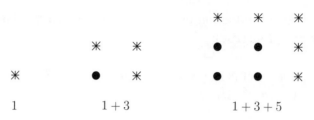

Figure 5.2. Sums of odd numbers

Because we assume (5.2) is true, we can add $N$ to both sides and the resulting equation will still be true. Thus

$$1 + 2 + 3 + \cdots + N - 1 + N = \frac{(N-1)N}{2} + N$$

$$= \frac{(N-1)N + 2N}{2} = \frac{N^2 - N + 2N}{2}$$

$$= \frac{N(N+1)}{2}.$$

In other words, under our induction hypothesis, $P_N$ is true. So we have checked (†) and we are finished. We have proven that $P_n$ is true for every positive integer $n$.

While we are adding first powers, we could try slightly fancier things, like skipping every other number:

$$1 + 3 + 5 + \cdots + (2n - 1) = n^2.$$

A heuristic proof of this may be seen in figure 5.2. You could also prove this by mathematical induction, which we leave to you as an exercise.

Adding even numbers doesn't give us anything new:

$$2 + 4 + 6 + \ldots + 2n = 2(1 + 2 + 3 + \cdots + n) = 2\left(\frac{n(n+1)}{2}\right) = n(n+1).$$

These numbers are neither triangular nor square, but nearly square, as in figure 5.3.

What if we skip two numbers each time?

$$1 + 4 + 7 + \cdots + (3n + 1) = \frac{3(n+1)^2 - (n+1)}{2}.$$

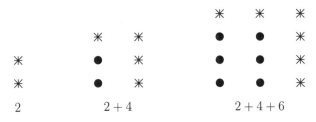

Figure 5.3. Sums of even integers

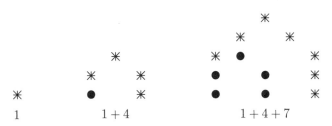

Figure 5.4. Pentagonal numbers

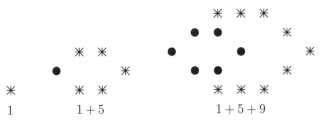

Figure 5.5. Hexagonal numbers

These are called pentagonal numbers. You can see the reason for the name in figure 5.4. Again, we leave the proof of this formula for you. You can also continue this game, creating hexagonal numbers (figure 5.5), septagonal numbers, and so on. These numbers are called *polygonal numbers*.

There's something interesting to point out here. The triangular numbers have the form $n(n+1)/2$, for positive $n$. However, you can substitute 0 or negative numbers for $n$. When you substitute 0, you get 0, and we can make 0 an honorary triangular number. When you substitute negative values for $n$, you get the same sequence as when you substitute positive values.

The squares have the form $n^2$. Again, there's nothing confusing about making 0 an honorary square, and again, substituting negative integer values for $n$ gives the same results as positive values.

The pentagonal numbers have the form $(3n^2 - n)/2$. For positive values of $n$, you get 1, 5, 12, 22, 35, .... Again, we can make 0 an honorary pentagonal number. But now negative values of $n$ yield the sequence 2, 7, 15, 26, 40, .... Notice that this is a completely different sequence of numbers. You can try to see if you can link this sequence with pentagons.

What about hexagonal numbers? The formula for these numbers is $2n^2 - n$, and the sequence is 1, 6, 15, 28, 45, .... Negative values of $n$ give the sequence 3, 10, 21, 36, 55, .... Is there any way to link this sequence with hexagons?[3]

We have seen elsewhere in this book that every positive integer is the sum of four squares. Cauchy proved that every positive integer is the sum of three triangular numbers, five pentagonal numbers, six hexagonal numbers, and so on. But this is not an easy theorem. Moreover, the largest number that is not a sum of at most four hexagonal numbers is 130.

What if we add consecutive squares together?

$$1^2 + 2^2 + 3^2 + \cdots + n^2 = ???$$

We were rather precipitate about nonmathematical induction in the physical sciences. Not only is it the way to go in natural science, but it has good uses in mathematics, too. On the one hand, if we have a statement $P_n$ that is true for all $n$ up to one billion, we would never accept this numerical verification as a proof that $P_n$ is true for all $n$. On the other hand, such a finding would lead us to try to prove that $P_n$ is true for all $n$. Sometimes not even a billion samples are needed.

Let's try to find a formula for the sum of the first $n$ squares and then let's try to prove it. We'll do some guessing. First, look back at the formulas we've already proven. The sum of $n$ 1's is $n$. The first $n$ first powers add to $\frac{n^2}{2} + \frac{n}{2}$. This is slim evidence, but we can

---

[3] We won't explore the topic of polygonal numbers any further, except to mention that the formula for the $n$th "$k$-gonal" number is always a *quadratic* polynomial in the variable $n$. The particular polynomial of course depends on $k$.

guess that the sum of the first $n$ $k$th powers will be a polynomial in $n$ of degree $k + 1$ with no constant term. If so, we are looking for a formula of the shape

$$1^2 + 2^2 + 3^2 + \cdots + n^2 = a_3 n^3 + a_2 n^2 + a_1 n.$$

It makes some sense that if we add up powers of a certain "dimension," we will get the next higher "dimension." And since the sum of zero powers is zero, it makes sense that there be no constant term.

We have three unknown coefficients: $a_1$, $a_2$, and $a_3$. We can try to find them with three equations. So let's work out what our conjectural formula says for $n = 1$, $2$, and $3$:

$$1 = a_3 + a_2 + a_1,$$

$$5 = 8a_3 + 4a_2 + 2a_1,$$

$$14 = 27a_3 + 9a_2 + 3a_1.$$

Here are three equations and three unknowns. We know how to solve them, although it is a little painful. Subtract 2 times the first equation from the second equation and 3 times the first equation from the third equation to get

$$3 = 6a_3 + 2a_2,$$

$$11 = 24a_3 + 6a_2.$$

Now, of these two equations in two unknowns, subtract 3 times the first from the second to obtain $2 = 6a_3$. Therefore $a_3 = \frac{1}{3}$. Substituting back in one of the pair of equations, we get $a_2 = \frac{1}{2}$. Substituting both these values into the first of the triplet of equations, we get $a_1 = \frac{1}{6}$. Putting these values back into $a_3 n^3 + a_2 n^2 + a_1 n$ and factoring, we find that the answer can be written simply as $\frac{n(n+1)(2n+1)}{6}$.

So we conjecture

$$1^2 + 2^2 + 3^2 + \cdots + n^2 = \frac{n(n + 1)(2n + 1)}{6}.$$

In fact, this is correct, and can be proved by mathematical induction. We will continue this line of thought in the next chapter.

# Chapter 6

# SUMS OF POWERS, USING LOTS OF ALGEBRA

## 1. History

It is one of the oldest stories in the book—but the books disagree on the details. Here's the version in an undergraduate textbook about the history of mathematics (Calinger, 1995):

> [T]he teacher, J.G. Büttner, asked students to sum the integers from 1 to 100. Gauss took only a second to write the answer on his slate.

Another source (Pólya, 1981) puts it differently:

> This happened when little Gauss still attended primary school. One day, the teacher gave a stiff task: To add up the numbers 1, 2, 3, and so on, up to 20. The teacher expected to have some time for himself while the boys were busy doing that long sum. Therefore, he was disagreeably surprised as the little Gauss stepped forward when the others had scarcely started working, put his slate on the teacher's desk, and said, "Here it is."

Pólya adds, "I particularly like [this] version which I heard as a boy myself, and I do not care whether it is authentic or not."

Here's another version (Bell, 1965):

> It was easy then for the heroic Büttner to give out a long problem in addition whose answer he could find by formula in a few seconds. The problem was of the following sort,

$81297 + 81495 + 81693 + \cdots + 1000899$, where the step from one number to the next is the same all along (here 198), and a given number of terms (here 100) are to be added.

It was the custom of the school for the boy who first got the answer to lay his slate on the table; the next laid his slate on top of the first, and so on. Büttner had barely finished stating the problem when Gauss flung his slate on the table: "There it lies," he said—"*Ligget se'*" in his peasant dialect.

The differing details do not affect the underlying mathematics, though they do cast doubt on the truth of the story. The usual explanation is as follows. We pick one version of the story and consider $1 + 2 + 3 + \cdots + 100$. Let $S$ be the sum of the numbers in question. Use basic properties of addition to write the sum in two different ways:

$$1 + \phantom{0}2 + \phantom{0}3 + \cdots + 98 + 99 + 100 = S,$$
$$100 + 99 + 98 + \cdots + \phantom{0}3 + \phantom{0}2 + \phantom{00}1 = S.$$

Add vertically, and the numbers in each column add to 101. There are 100 columns, so $2S = 100 \cdot 101$, or $S = 100 \cdot 101/2 = 5050$.

Replace 100 by any positive integer $n$, and the same argument shows that $S_1 = 1 + 2 + \cdots + n = n(n+1)/2$. For more details on adding consecutive integers, you can review chapter 5, section 2.

How could we compute the sum as given in Bell (1965)? The same trick works:

$$81297 + \phantom{0}81495 + \phantom{0}81693 + \cdots + 100899 = S,$$
$$100899 + 100701 + 100503 + \cdots + \phantom{0}81297 = S.$$

Add vertically, and each column sums to 182196. There are 100 columns, so $2S = 18219600$, and $S = 9109800$.

More generally, an *arithmetic series* is a sum of the form

$$a + (a + d) + (a + 2d) + \cdots + (a + (n-1)d).$$

The sum contains $n$ terms, and the difference between consecutive terms is $d$. If we reverse the order of the terms and add vertically, each column sums to $2a + (n-1)d$, and there are $n$ columns.

Therefore, the sum is $n(2a + (n - 1)d)/2$. Another way to derive this formula is to write $a + (a + d) + (a + 2d) + \cdots + (a + (n - 1)d) = na + d(1 + 2 + \cdots + (n - 1)) = na + dn(n - 1)/2$, using the result for $S_1$ in the previous paragraph. Our first example had $a = 1$, $d = 1$, and $n = 100$, and the sum is $100 \cdot 101/2$. Our second example had $a = 81297$, $d = 198$, and $n = 100$, and the sum is $100(182196)/2 = 9109800$ as before.

## 2. Squares

We now go further. A natural question[1] to consider next is to find a formula for $S' = 1^2 + 2^2 + 3^2 + \cdots + 100^2$. The same trick—writing $S' = 100^2 + 99^2 + \cdots + 2^2 + 1^2$—does not help this time, because different columns sum to different numbers: $100^2 + 1^2 \neq 99^2 + 2^2$.

One way to try to make progress is to rephrase our solution of the earlier problem in another guise. We can use summation notation to rephrase our trick. Let $S_1 = \sum_{i=0}^{n} i$. Starting the sum at 0 rather than 1 makes our life simpler, even though adding 0 to a sum does not change its value. Our reordering trick amounts to writing $S_1 = \sum_{i=0}^{n}(n - i)$. Adding, we get $2S_1 = \sum_{i=0}^{n} i + (n - i) = \sum_{i=0}^{n} n = n(n + 1)$, because there are $n + 1$ terms in the last sum, and all of the terms are equal to $n$. Finally, we get $S_1 = n(n + 1)/2$ as before.

We now can let

$$S_2 = \sum_{i=0}^{n} i^2 = \sum_{i=0}^{n}(n - i)^2.$$

---

[1] This question is particularly natural if you are trying to integrate $x^k$ and you do not know the Fundamental Theorem of Calculus. In other words, you are a mathematical contemporary of Pascal or Fermat. The connection with integration is that

$$\int_0^1 x^k \, dx = \lim_{n \to \infty} \frac{1^k + 2^k + 3^k + \cdots + n^k}{n^{k+1}},$$

which you can see by writing the integral as a limit of right-hand Riemann sums.

Then

$$2S_2 = \sum_{i=0}^{n} i^2 + (n-i)^2 = \sum_{i=0}^{n} i^2 + n^2 - 2ni + i^2$$

$$= \sum_{i=0}^{n} 2i^2 + n^2 - 2ni = \sum_{i=0}^{n} 2i^2 + \sum_{i=0}^{n} n^2 - 2n \sum_{i=0}^{n} i$$

$$= 2S_2 + (n+1)n^2 - 2nS_1.$$

Alas, the $2S_2$ term cancels on each side of the equation, and we are left with $(n+1)n^2 - 2nS_1 = 0$. This is true, but doesn't tell us anything we didn't already know.

At this point, we should remain optimistic because what went wrong suggests how to proceed. We did not get a formula for $S_2$, but we did get a different proof of the formula for $S_1$. Maybe trying to find a formula for $S_3 = 1^3 + 2^3 + \cdots + n^3$ in the same way will actually give a formula for $S_2$. We try

$$S_3 = \sum_{i=0}^{n} i^3 = \sum_{i=0}^{n} (n-i)^3,$$

$$2S_3 = \sum_{i=0}^{n} i^3 + (n-i)^3 = \sum_{i=0}^{n} i^3 + n^3 - 3n^2i + 3ni^2 - i^3$$

$$= \sum_{i=0}^{n} n^3 - 3n^2i + 3ni^2 = n^3(n+1) - 3n^2S_1 + 3nS_2$$

$$= n^3(n+1) - \tfrac{3}{2}n^3(n+1) + 3nS_2 = -\tfrac{1}{2}n^3(n+1) + 3nS_2,$$

and therefore

$$3nS_2 - 2S_3 = \tfrac{1}{2}n^3(n+1).$$

This is not looking promising, because we now have a formula relating *two* quantities we don't understand very well: $S_2$ and $S_3$.

But try once more, this time computing $S_4$:

$$S_4 = \sum_{i=0}^{n} i^4 = \sum_{i=0}^{n} (n-i)^4,$$

$$2S_4 = \sum_{i=0}^{n} i^4 + (n-i)^4 = \sum_{i=0}^{n} i^4 + n^4 - 4n^3 i + 6n^2 i^2 - 4n i^3 + i^4$$

$$= 2S_4 + n^4(n+1) - 4n^3 S_1 + 6n^2 S_2 - 4n S_3.$$

Cancel $2S_4$ from both sides of the equation, and substitute the known value of $S_1$:

$$0 = n^4(n+1) - 2n^4(n+1) + 6n^2 S_2 - 4n S_3$$

$$= -n^4(n+1) + 6n^2 S_2 - 4n S_3,$$

$$6n^2 S_2 - 4n S_3 = n^4(n+1).$$

We have, unfortunately, derived the same relationship between $S_2$ and $S_3$ a second time.

We need an additional trick. One method is to return to the formula for $S_3$, remove the $i = 0$ term, and use a summation trick called "reindexing":

$$S_3 = \sum_{i=0}^{n} i^3 = \sum_{i=1}^{n} i^3 = \sum_{i=0}^{n-1} (i+1)^3,$$

$$S_3 + (n+1)^3 = \sum_{i=0}^{n} (i+1)^3 = \sum_{i=0}^{n} i^3 + 3i^2 + 3i + 1$$

$$= S_3 + 3S_2 + 3S_1 + (n+1).$$

Now cancellation of $S_3$ from both sides and using our formula for $S_1$ is much more useful:

$$(n+1)^3 = 3S_2 + \frac{3}{2}n(n+1) + (n+1),$$

$$n^3 + 3n^2 + 3n + 1 = 3S_2 + \frac{3}{2}n^2 + \frac{5}{2}n + 1,$$

$$n^3 + \frac{3}{2}n^2 + \frac{1}{2}n = 3S_2,$$

$$\frac{n(n+1)(2n+1)}{2} = 3S_2,$$

$$\frac{n(n+1)(2n+1)}{6} = S_2.$$

We have a formula for $S_2$, and we can now derive a formula for $S_3$ from our relation $3nS_2 - 2S_3 = \frac{1}{2}n^3(n+1)$, but the derivation simultaneously is mathematically correct but aesthetically unsatisfying.

## 3. Divertimento: Double Sums

Making a problem more complicated sometimes turns out in the end to be a simplification. In the case of our sum $S_2$, the complication consists of writing a single sum as a double sum.

We know that $2^2 = 2+2$, $3^2 = 3+3+3$, and in general $i^2 = \overbrace{i+i+\cdots+i}^{i \text{ times}}$. Write this as $i^2 = \sum\limits_{j=1}^{i} i$, and we have

$$S_2 = \sum_{i=1}^{n} i^2 = \sum_{i=1}^{n}\sum_{j=1}^{i} i.$$

We now notice that if the second sum were slightly different, we could proceed. If only the second sum were $\sum\limits_{j=1}^{n} i$, it would add up to $in$, and then we would have

$$\sum_{i=1}^{n}\sum_{j=1}^{n} i = \sum_{i=1}^{n} in = n\sum_{i=1}^{n} i = n\frac{n(n+1)}{2}.$$

Let's turn this equation around:

$$n\frac{n(n+1)}{2} = \sum_{i=1}^{n}\sum_{j=1}^{n} i = \sum_{i=1}^{n}\sum_{j=1}^{i} i + \sum_{i=1}^{n}\sum_{j=i+1}^{n} i = S_2 + \sum_{i=1}^{n}\sum_{j=i+1}^{n} i.$$

For those who notice important details, we define the latter sum to be 0 when $i = n$, because we would have to let $j$ range from $n + 1$ to $n$.

How do we handle this last double sum? Notice that it is the sum of a bunch of 1's, then a smaller bunch of 2's, then a still smaller bunch of 3's, and so on:

$$
\begin{array}{rcc}
1 + 1 + 1 + 1 + \cdots + & 1 & + \\
2 + 2 + 2 + \cdots + & 2 & + \\
3 + 3 + \cdots + & 3 & + \\
4 + \cdots + & 4 & + \\
\cdots + & 5 & + \\
\cdots + & \cdots & + \\
& n - 1 &
\end{array}
$$

Adding vertically now gives us a way to make progress, because we can apply the formula that we already know. The first column gives $1 = \frac{1 \cdot 2}{2}$, the second gives $1 + 2 = \frac{2 \cdot 3}{2}$, the third gives $1 + 2 + 3 = \frac{3 \cdot 4}{2}$, all the way down to $1 + 2 + \cdots + (n - 1) = \frac{(n-1)n}{2}$. In other words,

$$
\sum_{i=1}^{n} \sum_{j=i+1}^{n} i = \sum_{k=1}^{n-1} \frac{k(k+1)}{2} = \sum_{k=1}^{n-1} \frac{k^2}{2} + \sum_{k=1}^{n-1} \frac{k}{2}
$$

$$
= \frac{1}{2} \sum_{k=1}^{n-1} k^2 + \frac{1}{2} \sum_{k=1}^{n-1} k = \frac{1}{2} \sum_{k=1}^{n-1} k^2 + \frac{(n-1)n}{4}.
$$

Finally, if we add and subtract $\frac{n^2}{2}$, we get

$$
\sum_{i=1}^{n} \sum_{j=i+1}^{n} i = \frac{1}{2} \sum_{k=1}^{n} k^2 + \frac{(n-1)n}{4} - \frac{n^2}{2} = \frac{S_2}{2} + \frac{(n-1)n}{4} - \frac{n^2}{2}.
$$

Putting the whole thing together,

$$
n \frac{n(n+1)}{2} = S_2 + \sum_{i=1}^{n} \sum_{j=i+1}^{n} i = S_2 + \frac{S_2}{2} + \frac{(n-1)n}{4} - \frac{n^2}{2}.
$$

We have an unavoidably messy bit of algebra:

$$\frac{3S_2}{2} = \frac{n^3 + n^2}{2} + \frac{n^2}{2} - \frac{n^2}{4} + \frac{n}{4} = \frac{n^3}{2} + \frac{3n^2}{4} + \frac{n}{4},$$

$$S_2 = \frac{n^3}{3} + \frac{n^2}{2} + \frac{n}{6} = \frac{2n^3 + 3n^2 + n}{6} = \frac{n(n+1)(2n+1)}{6}.$$

There's an alternative that we could have applied directly to our original double sum for $S_2$, called "interchanging the order of summation." We omit the details, but here is the computation if you wish to savor it:

$$S_2 = \sum_{i=1}^{n} i^2 = \sum_{i=1}^{n}\sum_{j=1}^{i} i = \sum_{j=1}^{n}\sum_{i=j}^{n} i = \sum_{j=1}^{n}\left[\frac{n(n+1)}{2} - \frac{(j-1)j}{2}\right]$$

$$= n\frac{n(n+1)}{2} - \sum_{j=1}^{n}\frac{j^2}{2} + \sum_{j=1}^{n}\frac{j}{2}$$

$$= n\frac{n(n+1)}{2} - \frac{S_2}{2} + \frac{n(n+1)}{4},$$

$$\frac{3S_2}{2} = n\frac{n(n+1)}{2} + \frac{n(n+1)}{4} = \frac{n^3}{2} + \frac{3n^2}{4} + \frac{n}{4},$$

with the same formula for $S_2$ after the same algebraic simplifications as before.

## 4. Telescoping Sums

These ideas are interesting, but it is not clear how to continue them for higher powers. Pascal found a systematic way to handle $S_k = 1^k + 2^k + \cdots + n^k$. His trick might seem unmotivated, but careful study shows the cleverness of this approach. To follow this, you need to remember the definition of binomial coefficients: $\binom{n}{k} = \frac{n!}{k!(n-k)!}$ for integers $k$ and $n$ with $0 \leq k \leq n$.

Use the binomial theorem to write $(x + 1)^k$ as $x^k + \binom{k}{1}x^{k-1} + \cdots + \binom{k}{k-1}x + 1$, and bring the first term on the right to the left:

$$(x + 1)^k - x^k = \binom{k}{1}x^{k-1} + \binom{k}{2}x^{k-2} + \cdots + \binom{k}{k-1}x + 1.$$

Now, let $x = 1, 2, 3, \ldots, n$, and sum this formula. The right-hand side gives

$$\binom{k}{1}S_{k-1} + \binom{k}{2}S_{k-2} + \cdots + \binom{k}{k-1}S_1 + n.$$

What about the left-hand side? We get $[2^k - 1^k] + [3^k - 2^k] + \cdots + [(n + 1)^k - n^k]$. Here we see one brilliant part of Pascal's idea: The left-hand side simplifies into $(n + 1)^k - 1$. (This is an example of a *telescoping sum*.) We get

$$(n + 1)^k = \binom{k}{1}S_{k-1} + \binom{k}{2}S_{k-2} + \cdots + \binom{k}{k-1}S_1 + (n + 1).$$

This formula allows us to compute $S_{k-1}$ for any $k$, but we need first to compute $S_1, S_2, \ldots, S_{k-2}$. Here's how it works. Pretend that we don't even know $S_1$. We apply the formula with $k = 2$ to get

$$(n + 1)^2 = \binom{2}{1}S_1 + (n + 1),$$

$$n^2 + 2n + 1 = 2S_1 + n + 1,$$

$$n^2 + n = 2S_1,$$

$$\frac{n^2 + n}{2} = S_1.$$

Now apply the formula with $k = 3$:

$$(n + 1)^3 = \binom{3}{2}S_2 + \binom{3}{1}S_1 + (n + 1),$$

$$n^3 + 3n^2 + 3n + 1 = 3S_2 + 3S_1 + n + 1$$

$$= 3S_2 + \frac{3n^2}{2} + \frac{3n}{2} + n + 1,$$

$$n^3 + \frac{3n^2}{2} + \frac{n}{2} = 3S_2.$$

This gives us the formula for $S_2$ yet one more time.

Perform one more step, just for fun. Apply the formula with $k = 4$:

$$(n + 1)^4 = \binom{4}{3}S_3 + \binom{4}{2}S_2 + \binom{4}{1}S_1 + (n + 1),$$

$$n^4 + 4n^3 + 6n^2 + 4n + 1 = 4S_3 + 6S_2 + 4S_1 + n + 1,$$

$$= 4S_3 + (2n^3 + 3n^2 + n) + (2n^2 + 2n) + n + 1,$$

$$n^4 + 2n^3 + n^2 = 4S_3,$$

$$\frac{n^2(n + 1)^2}{4} = S_3.$$

## 5. Telescoping Sums Redux

Pascal's telescoping summation idea is so clever that it cries out for further exploitation.[2] We have already summed the difference $(x + 1)^k - x^k$, but in our formulas the left-hand side of the sum will telescope for any function $f(x)$. We could sum $f(x + 1) - f(x)$ for $x = 1, 2, \ldots, n$, and get $f(n + 1) - f(1)$ to equal something. The question is how to choose the function $f(x)$ as cleverly as possible. Ideally, we want the "something" on the right-hand side to add up to $S_k$. The easiest way to arrange that is to find a function $f(x)$ such that $f(x + 1) - f(x) = x^k$. That way, when we add up the left-hand piece of our telescoping sum, we get $f(n + 1) - f(1)$. The right-hand side of the telescoping sum will give $1^k + 2^k \cdots + n^k = S_k$. If we are going to look for a function $f(x)$, we might as well be optimistic and search for a *polynomial* $p_k(x)$ such that $p_k(x + 1) - p_k(x) = x^k$. If we can find this elusive polynomial, then $S_k = p_k(n + 1) - p_k(1)$.

---

[2] WARNING: This section requires knowledge of the exponential function $e^x$ and elementary calculus, and it also uses some infinite series. You could skip it and return after reading chapters 7 and 8.

Such a polynomial does indeed exist, and it is defined in terms of *Bernoulli numbers* and *Bernoulli polynomials*. The fastest way to derive all of the properties is to make a definition that seems to be both unmotivated and hard to use. Take the function $te^{tx}/(e^t - 1)$, a function of the two variables $x$ and $t$, and expand it by powers of $t$. We make the definition:

$$\frac{te^{tx}}{e^t - 1} = \sum_{k=0}^{\infty} B_k(x)\frac{t^k}{k!} = B_0(x) + B_1(x)t + B_2(x)\frac{t^2}{2} + B_3(x)\frac{t^3}{6} + \cdots .$$

(6.1)

The functions on the right-hand side, $B_k(x)$, will be the Bernoulli polynomials. Unfortunately, we have no way of knowing just yet what they look like or how to compute them, or really whether this equation even makes any sense.

We can notice one thing from the definition (6.1). On the right-hand side, when we let $t \to 0$, we get the function $B_0(x)$. What about the left-hand side? Remember that the function $e^y$ has a nice series expansion:

$$e^y = \sum_{k=0}^{\infty} \frac{y^k}{k!} = 1 + y + \frac{y^2}{2} + \frac{y^3}{6} + \cdots .$$

So the left-hand side looks like

$$\frac{te^{tx}}{e^t - 1} = \frac{t\left(1 + (tx) + \frac{(tx)^2}{2} + \frac{(tx)^3}{6} + \cdots\right)}{t + \frac{t^2}{2} + \frac{t^3}{6} + \cdots} .$$

We can cancel a factor of $t$ from both the numerator and the denominator and get

$$\frac{te^{tx}}{e^t - 1} = \frac{\left(1 + (tx) + \frac{(tx)^2}{2} + \frac{(tx)^3}{6} + \cdots\right)}{1 + \frac{t}{2} + \frac{t^2}{6} + \cdots} .$$

Let $t \to 0$, and we see that this quotient tends to 1. We have deduced that $B_0(x)$ is the constant 1.

Before we compute any more of these functions, let's first see that the functions $B_k(x)$ more or less have the property we need. Remember that we're looking for a function $p_k(x)$, preferably a polynomial, so that $p_k(x + 1) - p_k(x) = x^k$. We can compute

$B_k(x+1) - B_k(x)$ by replacing $x$ with $x+1$ in (6.1), subtracting, and grouping:

$$\frac{t e^{t(x+1)}}{e^t - 1} = \sum_{k=0}^{\infty} B_k(x+1) \frac{t^k}{k!},$$

$$\frac{t e^{tx}}{e^t - 1} = \sum_{k=0}^{\infty} B_k(x) \frac{t^k}{k!},$$

$$\frac{t e^{t(x+1)} - t e^{tx}}{e^t - 1} = \sum_{k=0}^{\infty} \left( B_k(x+1) - B_k(x) \right) \frac{t^k}{k!}.$$

The most amazing thing happens to the left-hand side of this equation, which shows why we picked the mysterious function $t e^{tx}/(e^t - 1)$ in the first place:

$$\frac{t e^{t(x+1)} - t e^{tx}}{e^t - 1} = \frac{t e^{tx+t} - t e^{tx}}{e^t - 1} = \frac{t e^{tx}(e^t - 1)}{e^t - 1} = t e^{tx}.$$

We can plug in the series expansion for $e^{tx}$ and multiply by $t$:

$$\frac{t e^{t(x+1)} - t e^{tx}}{e^t - 1} = t \left( 1 + (tx) + \frac{(tx)^2}{2!} + \cdots \right)$$

$$= t + x t^2 + \frac{x^2 t^3}{2!} + \frac{x^3 t^4}{3!} + \cdots = \sum_{k=1}^{\infty} x^{k-1} \frac{t^k}{(k-1)!}.$$

When we compare coefficients of $t^k$, we get

$$\frac{B_k(x+1) - B_k(x)}{k!} = \frac{x^{k-1}}{(k-1)!}.$$

When we multiply through by $k!$, we get $B_k(x+1) - B_k(x) = k x^{k-1}$, for $k \geq 1$. Is that good enough? Yes. Replace $k$ by $k+1$, and divide by $k+1$ to get

$$\frac{B_{k+1}(x+1) - B_{k+1}(x)}{k+1} = x^k, \tag{6.2}$$

valid for $k > 0$. The polynomial we are seeking is $p_k(x) = B_{k+1}(x)/(k+1)$.

Now we can deduce a formula for $S_k$, the sum $1^k + 2^k + \cdots + n^k$:

$$S_k = \frac{B_{k+1}(n+1) - B_{k+1}(1)}{k+1}.$$  (6.3)

For example, when $k = 2$, we will see that $B_{k+1}(x) = x^3 - \frac{3}{2}x^2 + \frac{1}{2}x$, so

$$S_2 = \frac{B_3(n+1) - B_3(1)}{3}$$

$$= \frac{(n+1)^3 - \frac{3}{2}(n+1)^2 + \frac{1}{2}(n+1)}{3} = \frac{n^3}{3} + \frac{n^2}{2} + \frac{n}{6}.$$

So we need to figure out what this mysterious function $B_k(x)$ is in order to say that we have a satisfactory solution to the problem of sums of powers.

It is surprisingly easy to see that the functions $B_k(x)$ are actually polynomials. Return to (6.1), and differentiate both sides with respect to $x$. On the right-hand side, we get

$$\sum_{k=0}^{\infty} B_k'(x) \frac{t^k}{k!}.$$

On the left-hand side, we differentiate:

$$\frac{\partial}{\partial x} \left( \frac{te^{tx}}{e^t - 1} \right) = \frac{t^2 e^{tx}}{e^t - 1} = t \sum_{k=0}^{\infty} B_k(x) \frac{t^k}{k!}$$

$$= \sum_{k=0}^{\infty} B_k(x) \frac{t^{k+1}}{k!} = \sum_{k=1}^{\infty} B_{k-1}(x) \frac{t^k}{(k-1)!}.$$

When we compare the coefficients of $t^k$ on the left and right sides of the equation, we get

$$\frac{B_k'(x)}{k!} = \frac{B_{k-1}(x)}{(k-1)!}.$$

Now we multiply by $k!$ and get

$$B_k'(x) = k B_{k-1}(x).$$  (6.4)

Let's use (6.4) iteratively. We already worked out that $B_0(x)$ is the constant function 1. When $k = 1$, we get $B_1'(x) = B_0(x) = 1$. We can deduce that $B_1(x) = x + C$. We don't know yet what the constant of integration is (we'll learn soon enough), but it is customary to write the constant as $B_1$, the *first Bernoulli number*. We get $B_1(x) = x + B_1$. Now use (6.4) with $k = 2$: $B_2'(x) = 2x + 2B_1$. We get $B_2(x) = x^2 + 2B_1 x + C$, and the constant of integration is called $B_2$, the *second Bernoulli number*. If we do this again, we get $B_3'(x) = 3x^2 + 6B_1 x + 3B_2$, integration yields $B_3(x) = x^3 + 3B_1 x^2 + 3B_2 x + B_3$, and (you guessed it) $B_3$ is called the *third Bernoulli number*. By now, one piece of this pattern should be clear: Each of these functions $B_k(x)$ is actually a polynomial of degree $k$, which starts out $x^k$ and ends with the constant $B_k$. What happens in between those two terms is perhaps mysterious, but we can see that our goal is slightly closer: We have deduced that $B_k(x)$ is a polynomial of degree $k$.

We're not close to finished yet. We can substitute $x = 0$ into (6.2) and get $B_{k+1}(1) - B_{k+1}(0) = 0$ for $k > 0$. We can rephrase our result more usefully as

$$B_k(1) = B_k(0) \quad \text{if } k \geq 2. \tag{6.5}$$

Now we can compute the Bernoulli numbers. The equation $B_2(1) = B_2(0)$ says that $1 + 2B_1 + B_2 = B_2$, or $B_1 = -\frac{1}{2}$. Therefore, $B_1(x) = x - \frac{1}{2}$. The equation $B_3(1) = B_3(0)$ tells us that $1 + 3B_1 + 3B_2 + B_3 = B_3$, or $1 - \frac{3}{2} + 3B_2 = 0$, or $B_2 = \frac{1}{6}$. Now we know that $B_2(x) = x^2 - x + \frac{1}{6}$.

We can continue in this way, but there is at least one more symmetry in (6.1) that we have yet to exploit. We take that equation and simultaneously replace $t$ with $-t$ and $x$ with $1 - x$. The right-hand side becomes

$$\sum_{k=0}^{\infty} B_k(1-x) \frac{(-t)^k}{k!} = \sum_{k=0}^{\infty} (-1)^k B_k(1-x) \frac{t^k}{k!}. \tag{6.6}$$

The left-hand side of (6.1) undergoes a more interesting transformation,

$$\frac{(-t)e^{(-t)(1-x)}}{e^{-t} - 1} = \frac{te^{t(x-1)}}{1 - e^{-t}} = \frac{te^{tx}e^{-t}}{1 - e^{-t}},$$

and multiplication of the numerator and denominator by $e^t$ produces

$$\sum_{k=0}^{\infty}(-1)^k B_k(1-x)\frac{t^k}{k!} = \frac{te^{tx}}{e^t - 1} = \sum_{k=0}^{\infty} B_k(x)\frac{t^k}{k!}. \qquad (6.7)$$

Now, (6.7) tells us that $(-1)^k B_k(1-x) = B_k(x)$. If we substitute $x = 0$, we get $(-1)^k B_k(1) = B_k(0)$. If $k$ is even, this is just (6.5). But when $k$ is odd and at least 3, we deduce that $-B_k(1) = B_k(0) = B_k(1)$, meaning that $B_k(1) = 0$, and hence $B_k(0) = 0$. In other words, $B_3 = B_5 = B_7 = \cdots = 0$. We can now compute $B_3(x)$, because $B_3'(x) = 3B_2(x)$ and the constant of integration is 0; we get $B_3(x) = x^3 - \frac{3}{2}x^2 + \frac{1}{2}x$.

Remember that each Bernoulli polynomial can be computed (up to the constant of integration, which is $B_k$ by definition) by integration. That gives us a formula for the Bernoulli polynomials in terms of the Bernoulli numbers:

$$B_k(x) = x^k + \binom{k}{1}B_1 x^{k-1} + \binom{k}{2}B_2 x^{k-2} + \cdots + \binom{k}{k-1}B_{k-1}x + B_k.$$

How can we prove this formula? We use induction on $k$. First, we check it when $k = 1$. The left-hand side is just $B_1(x)$, and the right-hand side is $x + B_1 = x - \frac{1}{2}$, which agrees with our computation earlier. The only other thing that we need to verify is that $B_k'(x) = kB_{k-1}(x)$. We take the right-hand side and differentiate:

$$B_k'(x) = \left(x^k + \binom{k}{1}B_1 x^{k-1} + \binom{k}{2}B_2 x^{k-2} + \cdots + \binom{k}{k-1}B_{k-1}x + B_k\right)'$$

$$= kx^{k-1} + (k-1)\binom{k}{1}B_1 x^{k-2} + (k-2)\binom{k}{2}B_2 x^{k-3}$$

$$+ \cdots + \binom{k}{k-1}B_{k-1}.$$

Now there's a nice identity to apply:

$$(k-j)\binom{k}{j} = (k-j)\frac{k!}{j!(k-j)!} = \frac{k!}{j!(k-j-1)!}$$

$$= k\frac{(k-1)!}{j!(k-j-1)!} = k\binom{k-1}{j}.$$

TABLE 6.1. Bernoulli Numbers

| $k$ | 0 | 1 | 2 | 4 | 6 | 8 | 10 | 12 | 14 | 16 | 18 |
|---|---|---|---|---|---|---|---|---|---|---|---|
| $B_k$ | 1 | $-\frac{1}{2}$ | $\frac{1}{6}$ | $-\frac{1}{30}$ | $\frac{1}{42}$ | $-\frac{1}{30}$ | $\frac{5}{66}$ | $-\frac{691}{2730}$ | $\frac{7}{6}$ | $-\frac{3617}{510}$ | $\frac{43867}{798}$ |

That gives

$$B_k'(x) = k\left(x^{k-1} + \binom{k-1}{1}B_1 x^{k-2} + \binom{k-1}{2}B_2 x^{k-3}\right.$$

$$\left. + \cdots + \binom{k-1}{k-1}B_{k-1}\right)$$

$$= kB_{k-1}(x)$$

by the inductive hypothesis. We have now demonstrated

$$B_k(x) = x^k + \binom{k}{1}B_1 x^{k-1} + \binom{k}{2}B_2 x^{k-2} + \cdots + \binom{k}{k-1}B_{k-1}x + B_k.$$

Use the formula $B_k(1) = B_k(0) = B_k$ if $k \geq 2$, set $x = 1$ and subtract $B_k$ from both sides of the equation, and get

$$1 + \binom{k}{1}B_1 + \binom{k}{2}B_2 + \cdots + \binom{k}{k-1}B_{k-1} = 0, \quad \text{if } k \geq 2. \quad (6.8)$$

Equation (6.8) is actually the standard way to define and compute the Bernoulli numbers inductively. Start with $k = 2$, and we get $1 + 2B_1 = 0$, or $B_1 = -\frac{1}{2}$. Next take $k = 3$, and we get $1 + 3B_1 + 3B_2 = 0$, giving $B_2 = \frac{1}{6}$. Next take $k = 4$, and we get $1 + 4B_1 + 6B_2 + 4B_3 = 0$, giving (as expected) $B_3 = 0$. You can continue computing Bernoulli numbers for as long as you like. We list a few of them in table 6.1. Note how the values jump around in a mysterious way. In table 6.2, we list a few values of the polynomial $S_k$. Remember that $S_k = 1^k + 2^k + 3^k + \cdots + n^k$, the motivation for

TABLE 6.2. $S_k$

| $k$ | $S_k$ |
|---|---|
| 1 | $n(n+1)/2$ |
| 2 | $n(n+1)(2n+1)/6$ |
| 3 | $n^2(n+1)^2/4$ |
| 4 | $n(n+1)(2n+1)(3n^2+3n-1)/30$ |
| 5 | $n^2(n+1)^2(2n^2+2n-1)/12$ |
| 6 | $n(n+1)(2n+1)(3n^4+6n^3-3n+1)/42$ |
| 7 | $n^2(n+1)^2(3n^4+6n^3-n^2-4n+2)/24$ |

this entire exploration, and the polynomials in table 6.2 come from (6.3).

We close with one more remark. Because the Bernoulli numbers are the constants in the Bernoulli polynomials, we can set $x = 0$ in (6.1):

$$\frac{t}{e^t - 1} = \sum_{k=0}^{\infty} B_k \frac{t^k}{k!} = B_0 + B_1 t + B_2 \frac{t^2}{2} + B_3 \frac{t^3}{6} + \cdots.$$

We can now use the theory of Taylor series to deduce that if $f(t) = t/(e^t - 1)$, then $f^{(k)}(0) = B_k$.

## 6. Digression: Euler–Maclaurin Summation

There is a beautiful application of these ideas to the problem of approximating integrals by sums or vice versa. Suppose that $f(x)$ is a function that can be differentiated arbitrarily many times. We will start by estimating $\int_0^1 f(x)\,dx$. We know that $B_0(x) = 1$, so we can include $B_0(x)$ as a factor into our integral. Now, we can integrate by parts, setting $u = f(x)$, $du = f'(x)\,dx$, $dv = B_0(x)\,dx$, and $v = B_1(x)$. Remember that $B_1(x) = x - \frac{1}{2}$, so that $B_1(1) = \frac{1}{2}$ and $B_1(0) = -\frac{1}{2}$.

We get

$$\int_0^1 f(x)\,dx = \int_0^1 B_0(x)f(x)\,dx = B_1(x)f(x)\Big|_0^1 - \int_0^1 B_1(x)f'(x)\,dx$$

$$= \frac{1}{2}(f(1)+f(0)) - \int_0^1 B_1(x)f'(x)\,dx.$$

Notice that we can interpret this formula by saying that the area under the curve $y = f(x)$ from $x = 0$ to $x = 1$ is approximated by what might be our first guess: the average of the heights of the curve at $x = 0$ and $x = 1$, with the error term given by the last integral.

Now we integrate by parts again, setting $u = f'(x)$, $du = f''(x)\,dx$, $dv = B_1(x)\,dx$, and $v = B_2(x)/2$. We get

$$\int_0^1 f(x)\,dx = \frac{1}{2}(f(1)+f(0)) - \frac{B_2(x)f'(x)}{2}\Big|_0^1 + \int_0^1 \frac{B_2(x)}{2}f''(x)\,dx.$$

Because $B_2(1) = B_2(0) = B_2$, this simplifies to

$$\int_0^1 f(x)\,dx = \frac{1}{2}(f(1)+f(0)) - \frac{B_2}{2}(f'(1)-f'(0)) + \int_0^1 \frac{B_2(x)}{2}f''(x)\,dx.$$

It is after some more integration by parts that we see why this is a good idea. We set $u = f^{(2)}(x)$, $du = f^{(3)}(x)\,dx$, $dv = B_2(x)/2\,dx$, and $v = B_3(x)/3!$. (Remember that $f^{(m)}(x)$ denotes the $m$th derivative of the function $f(x)$.) Because $B_3(0) = B_3(1) = 0$, we just get

$$\int_0^1 f(x)\,dx = \frac{1}{2}(f(1)+f(0)) - \frac{B_2}{2}(f'(1)-f'(0)) - \int_0^1 \frac{B_3(x)}{3!}f^{(3)}(x)\,dx.$$

Integrate by parts again, setting $u = f^{(3)}(x)$, $du = f^{(4)}(x)\,dx$, $dv = B_3(x)/3!\,dx$, and $v = B_4(x)/4!$, and remembering that

$B_4(1) = B_4(0) = B_4$:

$$\int_0^1 f(x)\,dx = \frac{1}{2}(f(1) + f(0)) - \frac{B_2}{2}\left(f'(1) - f'(0)\right)$$

$$-\frac{B_4(x)}{4!}f^{(3)}(x)\Big|_0^1 + \int_0^1 \frac{B_4(x)}{4!}f^{(4)}(x)\,dx$$

$$= \frac{1}{2}(f(1) + f(0)) - \frac{B_2}{2}\left(f'(1) - f'(0)\right)$$

$$-\frac{B_4}{4!}\left(f^{(3)}(1) - f^{(3)}(0)\right) + \int_0^1 \frac{B_4(x)}{4!}f^{(4)}(x)\,dx.$$

After many repetitions, we get

$$\int_0^1 f(x)\,dx = \frac{1}{2}(f(1) + f(0)) - \sum_{r=1}^{k} \frac{B_{2r}}{(2r)!}\left(f^{(2r-1)}(1) - f^{(2r-1)}(0)\right)$$

$$+ \int_0^1 \frac{B_{2k}(x)}{(2k)!}f^{(2k)}(x)\,dx. \tag{6.9}$$

One application of (6.9) is that for many functions $f(x)$, the integral on the right-hand side of the formula is very small (because $(2k)!$ gets to be very big), and so we can approximate the integral on the left-hand side with the other terms on the right-hand side. For example, we can apply the formula with $f(x) = \cos x$. In that case, we know the value of the integral: $\int_0^1 \cos x\,dx = \sin 1 - \sin 0 \approx 0.8415$. The first term on the right-hand side of (6.9) is just the trapezoidal approximation to the integral. That gives 0.7702. The next term gives $-\frac{B_2}{2}(-\sin 1 + \sin 0) \approx 0.0701$, and just those two terms sum to 0.8403. The next term gives $-\frac{B_4}{24}(\sin 1 - \sin 0) \approx 0.0012$, and we now have the correct answer to nearly four decimal places.

Typically, one applies (6.9) by sliding the interval of integration from $[0,1]$ to $[1,2]$ to $[2,3]$ and so on up to $[n-1, n]$, and then summing the resulting equations. All of the terms are easy to sum except for the integral on the right-hand side. The "main term" gives the trapezoidal approximation to the integral again, and the other sums are telescoping sums. If we cheat and just call the sum of the integrals $R_k(n)$ for "remainder," we get

$$\int_0^n f(x)\,dx = \frac{1}{2}f(0) + f(1) + \cdots + f(n-1) + \frac{1}{2}f(n)$$

$$- \sum_{r=1}^{k} \frac{B_{2r}}{(2r)!}\left(f^{(2r-1)}(n) - f^{(2r-1)}(0)\right) + R_k(n).$$

Calculus books that cover Euler–Maclaurin summation contain information about how to estimate $R_k(n)$ in particular situations.

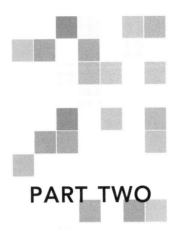

# PART TWO

# Infinite Sums

*Chapter 7*

# INFINITE SERIES

From now on, we assume that you are familiar with the concept of limit from calculus. Although the subject of this chapter is infinite series, which refers to sums of infinitely many numbers or terms, we will begin with a finite sum of a particular sort.

## 1. Finite Geometric Series

If we add up some consecutive powers of $a$, we get a *geometric series*. In greatest generality, a finite geometric series looks like

$$ca^m + ca^{m+1} + \cdots + ca^n$$

with nonzero constants $a$ and $c$ and integers $m \leq n$. We say this geometric series "has ratio $a$." There is a nice formula for this sum:

$$ca^m + ca^{m+1} + \cdots + ca^n = \frac{ca^m - ca^{n+1}}{1 - a}. \qquad (7.1)$$

One way to remember this formula is to say that the sum of a finite geometric series equals the first term minus "the term that is one past the last term," all divided by one minus the ratio. This formula only works if $a \neq 1$. (When $a = 1$, you don't really need any formula.)

Why is the formula true? Let

$$S = 1 + a^1 + a^2 \cdots + a^{n-m}.$$

Then

$$aS = a^1 + a^2 + a^3 + \cdots + a^{n-m+1}.$$

Subtract $aS$ from $S$, and all of the inside terms cancel. We get

$$S - aS = 1 - a^{n-m+1}.$$

But we can rewrite the left-hand side of this equation as

$$S - aS = S(1 - a).$$

Now, divide both sides by $1 - a$, and we get

$$1 + a^1 + a^2 + \cdots + a^{n-m} = \frac{1 - a^{n-m+1}}{1 - a}. \qquad (7.2)$$

Multiply both sides by $ca^m$, and you get (7.1).

For example, we can easily sum up the powers of 2:

$$1 + 2 + 4 + 8 + 16 + \cdots + 2^n = \frac{1 - 2^{n+1}}{1 - 2} = 2^{n+1} - 1.$$

So the sum is one less than the next power of 2. Notice that this sum grows without bound as $n$ increases.

We can just as easily sum up the powers of $\frac{1}{2}$:

$$\frac{1}{2} + \frac{1}{4} + \cdots + \frac{1}{2^n} = \frac{\frac{1}{2} - \frac{1}{2^{n+1}}}{1 - \frac{1}{2}} = 1 - \frac{1}{2^n}.$$

This sum does have a limit as $n$ increases, namely 1. This limiting behavior is related to one of Zeno's paradoxes. If you try to exit a room, first you have to go halfway to the door, then half of the remaining distance (which is a distance of one-quarter of the way to the door), and so on. Thus you can never actually get out of the room. Our sum shows that after $n$ "steps" of this "process," you will still have a distance to go of $1/2^n$ of the distance you had to begin with. The "solution" to the paradox is that this is an infinite process that you can actually complete by letting $n$ go to infinity, at which time you will have gone the whole distance, since $\lim_{n \to \infty} 1 - \frac{1}{2^n} = 1$.

(We are not claiming that this is a satisfactory resolution of the paradox from a philosophical point of view or that the paradox actually makes sense from a physical point of view.)

Our formula (7.1) works also for negative ratios. An interesting case is $a = -1$. Then we have the formula $1 - 1 + 1 - 1 + 1 \cdots + (-1)^n = (1 - (-1)^{n+1})/(1 - (-1))$, which equals $0/2 = 0$ if $n$ is odd and $2/2 = 1$ if $n$ is even. (Our series starts with exponent 0. Note that $(-1)^0 = 1$.) What happens now if $n$ goes to infinity? We get oscillating behavior.[1]

We haven't discussed complex numbers just yet, but in fact our formula works for any complex number not equal to 1. You can try it for yourself. Try, for example, $a = 1 + i$ or $a = i$. (BETTER: Try both.)

## 2. Infinite Geometric Series

Suppose now that we try to add up a geometric series with an infinite number of terms. Let $a$ be any nonzero real number, and consider the series

$$1 + a + a^2 + a^3 + \cdots .$$

If you want to consider the most general geometric series, just multiply through by a constant.

We have seen in equation (7.2), with $m = 1$, that the sum of the first $n$ terms of this series is

$$s_n = 1 + a + a^2 + a^3 + \cdots + a^{n-1} = \frac{1 - a^n}{1 - a}.$$

We see immediately that these sums $s_n$ have a limit as $n \to \infty$ if the absolute value $|a|$ is less than 1. In this case, $\lim_{n \to \infty} a^n = 0$, and we obtain $\lim_{n \to \infty} s_n = \frac{1}{1-a}$. We say that the geometric series converges[2] to $\frac{1}{1-a}$. If $|a| \geq 1$, then the sums $s_n$ do not have a limit. We can summarize our reasoning in the following very important theorem:

---

[1] See section 2 of the Introduction for a different interpretation of what "happens" when $n \to \infty$.

[2] We say that a series *converges* when the sequence of partial sums has a limit. We explain this more carefully in section 6 of this chapter.

**THEOREM 7.3**: *The series* $1 + a + a^2 + a^3 + \cdots$ *converges if and only if* $|a| < 1$, *in which case it converges to* $\frac{1}{1-a}$.

A significant psychological step is taken by reversing things. We can say the fraction $\frac{1}{1-a}$ can be "developed" into the power series $1 + a + a^2 + a^3 + \cdots$. We will explore this reversal later by replacing $a$ by a variable.

Going back to the example of Zeno's paradox in the previous section, we see theorem 7.3 agrees with what we said there. Because $\frac{1}{2}$ has absolute value less than 1, the theorem applies and we get

$$\frac{1}{2} + \frac{1}{4} + \frac{1}{8} + \cdots = \frac{1}{2}\left(1 + \frac{1}{2} + \left(\frac{1}{2}\right)^2 + \cdots\right) = \frac{1}{2}\left(\frac{1}{1 - \frac{1}{2}}\right) = 1.$$

The answer 1 means that you do indeed get to the doorway leading out of the room.

Another example is the familiar fact that the infinite repeating decimal $0.999999\ldots$ equals 1. This assertion is often unsettling to people the first time they see it, because they may not understand that the meaning of an infinite decimal has to be given in terms of a limit. We have seen students and adults refuse to believe this fact, essentially because they are stuck in a kind of mental version of Zeno's paradox. Why is this fact true? From the definition of decimal notation,

$$0.999999\ldots = .9 + .09 + .009 + \cdots$$

$$= .9\left(1 + \frac{1}{10} + \left(\frac{1}{10}\right)^2 + \cdots\right) = (.9)\frac{1}{1 - \frac{1}{10}} = (.9)\left(\frac{10}{9}\right) = 1.$$

### 3. The Binomial Series

The binomial theorem may be the second most famous result in mathematics, lagging only the Pythagorean theorem in its fame. One reference in popular culture can be found in "The Final

Problem," by Sir Arthur Conan Doyle, in which Sherlock Holmes states:

> [Professor Moriarty] is a man of good birth and excellent ed-
> ucation, endowed by nature with a phenomenal mathematical
> faculty. At the age of twenty-one he wrote a treatise upon
> the Binomial Theorem, which has had a European vogue. On
> the strength of it he won the Mathematical Chair at one of
> our smaller universities, and had, to all appearances, a most
> brilliant career before him.

It is hard to imagine what Moriarty could have discovered, because in actuality the theory was understood well before Doyle began writing about Holmes.

If $n$ is a positive integer, the expansion of $(1 + x)^n$ was discovered separately by many classical cultures:

$$(1 + x)^n = 1 + \binom{n}{1}x + \binom{n}{2}x^2 + \cdots + \binom{n}{n}x^n.$$

This equation is usually taught to high school students, and we used it already in chapter 6. Newton discovered a generalization in which the exponent $n$ can be replaced by any real number. Recall that the numbers $\binom{n}{r}$ are called *binomial coefficients*. The definition is

$$\binom{n}{r} = \frac{n!}{r!(n - r)!}, \tag{7.4}$$

provided that $r$ and $n$ are nonnegative integers, with $0 \leq r \leq n$. (Remember that $0!$ is defined to be 1, which allows us to define a binomial coefficient when $r = 0$ or $r = n$.)

The interesting problem is to extend the definition of $\binom{n}{r}$ by replacing $n$ by any real number $\alpha$. The generalization is easy to explain, though finding it took considerable insight. We can rewrite (7.4) by canceling matching factors in the numerator and denominator to get

$$\binom{n}{r} = \frac{n!}{r!(n-r)!}$$

$$= \frac{n(n-1)(n-2)\cdots(n-r+1)(n-r)!}{r!(n-r)!}$$

$$= \frac{n(n-1)(n-2)\cdots(n-r+1)}{r!}.$$

Now we have our generalization. If $\alpha$ is any real number and $r$ is any positive integer, we define

$$\binom{\alpha}{r} = \frac{\alpha(\alpha-1)(\alpha-2)\cdots(\alpha-r+1)}{r!}.$$

For example,

$$\binom{\frac{1}{3}}{4} = \frac{\left(\frac{1}{3}\right)\cdot\left(\frac{-2}{3}\right)\cdot\left(\frac{-5}{3}\right)\cdot\left(\frac{-8}{3}\right)}{24} = -\frac{10}{243}.$$

We extend the definition to define $\binom{\alpha}{0}$ to be 1, just as $\binom{n}{0} = 1$.

With this definition, the generalization follows immediately:

**THEOREM 7.5**: *If $|x| < 1$ and $\alpha$ is any real number, then*

$$(1+x)^\alpha = 1 + \binom{\alpha}{1}x + \binom{\alpha}{2}x^2 + \binom{\alpha}{3}x^3 + \binom{\alpha}{4}x^4 + \cdots$$

$$= 1 + \alpha x + \frac{\alpha(\alpha-1)}{2}x^2 + \frac{\alpha(\alpha-1)(\alpha-2)}{6}x^3$$

$$+ \frac{\alpha(\alpha-1)(\alpha-2)(\alpha-3)}{24}x^4 + \cdots. \qquad (7.6)$$

For example,

$$(1+x)^{1/3} = 1 + \frac{1}{3}x - \frac{1}{9}x^2 + \frac{5}{81}x^3 - \frac{10}{243}x^4 + \cdots. \qquad (7.7)$$

Notice that theorem 7.5 generalizes theorem 7.3. Start with $(1+x)^\alpha$, replace $x$ with $-a$, and set $\alpha = -1$, and we get an

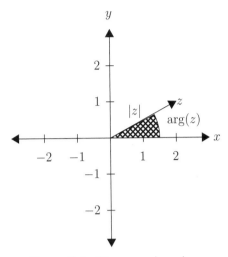

Figure 7.1. The complex plane

alternative formula for the infinite geometric series:

$$(1-a)^{-1} = 1 - \binom{-1}{1}a + \binom{-1}{2}a^2 - \binom{-1}{3}a^3 + \cdots .$$

You can convince yourself that $\binom{-1}{r} = (-1)^r$, and thereby verify that the binomial theorem is consistent with the formula for the sum of an infinite geometric series.

## 4. Complex Numbers and Functions

Our theory gets more interesting when we allow the use of complex numbers. A brief review of the basic facts about complex numbers is in order. As usual, let's use $x$ and $y$ for Cartesian coordinates in the plane, so that $x$ and $y$ are arbitrary real numbers. We can draw axes as in figure 7.1.

We can interpret this picture in a different way as the complex plane. It is traditional to use $z$ for a generic complex variable and to write $z = x + iy$. So $x$ is the *real part* of $z$, and $y$ is the *imaginary part* of $z$. We write $x = \text{Re}(z)$ and $y = \text{Im}(z)$. Complex numbers and the geometry of the complex plane are discussed in lots of textbooks. You could also review them in Ash and Gross (2006, chapter 5; 2012, chapter 2, section 4). Barry Mazur wrote a very interesting book (Mazur, 2003) about the philosophy and poetics of complex numbers.

The complex number $z = x + iy$ is "identified" with the point in the plane with Cartesian coordinates $(x, y)$. The *norm* (also called the absolute value) of a complex number $z$ is written $|z|$ and equals $\sqrt{x^2 + y^2}$ (by convention, the square-root sign applied to a positive number designates the positive square root). This is the same as the distance from $z$ to $0 = 0 + 0i$ in the plane. If you draw a line segment from $0$ to $z$ and note the angle it makes moving counterclockwise from the positive $x$-axis, then this angle (measured in radians) is called the *argument* of $z$ and is written $\arg(z)$. The definition of the tangent tells you that $\tan(\arg(z)) = \frac{y}{x}$. The argument of $0$ is not defined.

The norm of $z$ is always a nonnegative real number, and it is positive if $z \neq 0$. By convention, we usually take the argument of $z$ to be nonnegative and less than $2\pi$. A glance at the diagram will show you that any complex number is determined if you know both its norm and its argument (or just its norm if the norm is 0). The norm and argument give you another kind of coordinate pair for $z$, called *polar coordinates*. The norm and argument behave nicely under multiplication: $|zw| = |z||w|$ and $\arg(zw) \equiv \arg(z) + \arg(w)$ (mod $2\pi$) for any two complex numbers $z$ and $w$. (Here we extend the congruence notation. We write $a \equiv b$ (mod $2\pi$) to mean that $\frac{b-a}{2\pi}$ is an integer.)

The whole complex plane we call $\mathbf{C}$, which is also the name for the field of complex numbers. That dual use is not a problem, because we are "identifying" each complex number $z = x + iy$ with the point in the plane that has coordinates $(x, y)$.

Now that we have defined the complex plane $\mathbf{C}$, we can think about functions $f : \mathbf{C} \to \mathbf{C}$ that take a complex number as input and output a complex number. Often, we need to use functions $f : A \to \mathbf{C}$ that are only defined for a subset $A$ of $\mathbf{C}$. For our purposes in this context, we will always require $A$ to be an "open set."

**DEFINITION**: An *open set* in the complex plane is a subset $\Omega$ of the plane with the property that if $\Omega$ contains a point $a$, then $\Omega$ also contains some open disc, possibly very small, centered at $a$.

In particular, the whole plane is an open set. Other important examples of open sets are the open unit disc $\Delta^0$ and the upper half-plane $H$, both of which are defined later.

There is a class of complex functions of huge importance called "analytic functions."

**DEFINITION**: If $\Omega$ is an open set and $f : \Omega \to \mathbf{C}$ is a function, we say that $f(z)$ is an *analytic function* of $z$ on $\Omega$ if $f(z)$ can be differentiated with respect to $z$ at every point in $\Omega$. Namely, the limit

$$\lim_{h \to 0} \frac{f(z_0 + h) - f(z_0)}{h}$$

exists for all $z_0$ in $\Omega$. When this limit exists, we say that $f(z)$ is *complexly differentiable* at $z = z_0$, and the limit is called the complex derivative $f'(z_0)$. Moreover, for any particular $z_0$, we require that the limit be the same, no matter how the complex number $h$ tends to 0. (In the complex plane, there are many different directions in which $h$ can tend to 0, and the limit has to be the same number for each direction.)

For example, any polynomial in $z$ is analytic on the whole plane, and a theorem states that any convergent power series (see section 8 of this chapter) in $z$ is analytic within its radius of convergence. Being analytic is a rather stringent condition on functions.

## 5. Infinite Geometric Series Again

Return to our infinite geometric series, with ratio $a$, and let $a$ vary over the set of complex numbers $\mathbf{C}$. Because that will turn $a$ into a complex variable, it is traditional to use the letter $z$ instead. We denote by $\Delta^0$ the set of all complex numbers of norm $< 1$:

$$\Delta^0 = \{z = x + iy \mid x^2 + y^2 < 1\}.$$

We call $\Delta^0$ the *open unit disc*. There is one critical observation that allows us to proceed from real values to complex ones. If $|z| < 1$ (i.e., if $z$ is an element of $\Delta^0$), then $\lim_{k \to \infty} z^k = 0$. (Quick proof: If $|z| < 1$, then $\lim_{k \to \infty} |z|^k = 0$. Therefore $\lim_{k \to \infty} |z^k| = 0$, which implies that $\lim_{k \to \infty} z^k = 0$.) Now we can rephrase theorem 7.3 to say:

**THEOREM 7.8:** *The function $\frac{1}{1-z}$ can be developed into the infinite geometric series $1 + z + z^2 + \cdots$ in the open unit disc. That is, for any complex number $z$ with $|z| < 1$, we have the equality*

$$\frac{1}{1-z} = 1 + z + z^2 + \cdots .$$

We call this "the formula for the infinite geometric series."

It may seem like we haven't said anything new. In fact, we haven't. But psychologically, we are now talking about two *functions* of $z$ that are equal for all $z$ in a certain domain. It is even worth giving a calculus proof of this formula. We recognize that the right-hand side looks like a Taylor series around $z = 0$. So if this formula is correct, it had better be the Taylor series of the left-hand side.

The rule for finding the Taylor series about $z = 0$ for an analytic complex function $f(z)$ with 0 in its domain is the same as the rule for Taylor series of a function of a real variable. (If the presence of complex numbers is troubling you, just change $z$ to $x$ and do the usual Taylor series derivation.) The Taylor series of $f(z)$ about $z = 0$ is the series

$$a_0 + a_1 z + a_2 z^2 + \cdots ,$$

where $a_0 = f(0)$, $a_1 = f'(0)$, $a_2 = \frac{1}{2!}f''(0), \ldots, a_n = \frac{1}{n!}f^{(n)}(0), \ldots$. Here, $f^{(n)}(z)$ is the $n$th complex derivative of $f$ with respect to $z$.

Using the chain rule and the power rule, we can find the derivatives of

$$f(z) = \frac{1}{1-z} = (1-z)^{-1}.$$

We obtain

$$f'(z) = (1-z)^{-2},$$

$$f''(z) = 2(1-z)^{-3},$$

and so on. The result is

$$f^{(n)}(z) = (n!)(1-z)^{-n-1}.$$

Now, plug in $z = 0$ to obtain the general formula (a very nice one)

$$f^{(n)}(0) = n!$$

for all $n$. It follows that the coefficients $a_n$ of the Taylor series of $f(z)$ are all equal to 1. We have again proved the formula for the infinite geometric series in $\Delta^0$.

This method of proof can be generalized to prove (7.6), the infinite binomial series.

## 6. Examples of Infinite Sums

We now want to consider more general forms of infinite sums than just geometric series. We begin with the usual definition: An infinite series is an infinite sum of terms like

$$a_1 + a_2 + a_3 + \cdots.$$

If the "partial sums" $a_1 + a_2 + a_3 + \cdots + a_n$ tend to a limit $L$ as $n$ goes to infinity, then we say the series *converges*. We say it converges to $L$, or has the limit $L$. (Remember that converging to $L$ means getting closer and closer to $L$ as $n$ gets larger, and getting as close as we like and staying at least that close from some $N$ onward.) From the definition of convergence, it follows that if an infinite series converges, the individual terms $a_n$ must tend to zero as $n$ goes to infinity—otherwise, the partial sums wouldn't stay as close as we like to $L$ from some point onward.[3]

Here are some examples of convergent infinite series. The first is any infinite decimal number. Just as an integer written in base 10 is secretly a problem in addition (see chapter 5, section 1), so any infinite decimal is secretly an infinite series.

For example, by definition of base 10 notation,

$$1.23461\ldots = 1 + (2 \times 10^{-1}) + (3 \times 10^{-2}) + (4 \times 10^{-3})$$
$$+ (6 \times 10^{-4}) + (1 \times 10^{-5}) + \cdots.$$

How can we see that this series converges? If we want the partial sums to be within $10^{-14}$, say, of the limit, we only have to ensure that we take at least 16 terms. That's because the remaining terms add up to an amount less than or equal to $.999999\ldots \times 10^{-15} = 10^{-15}$, which is surely less than $10^{-14}$. (The fact that the limit exists is due to the "completeness axiom" for the real numbers, which asserts basically that there are no "holes" in the real number line.)

The decimal is eventually repeating if and only if the number it represents is the ratio of two integers (i.e., it is a "rational number"). This is proven in all textbooks that deal with infinite decimals. Shall we review the proof here? Suppose the real number $b$ is represented by an infinite, eventually repeating decimal, and the repeating pattern has length $k$. Multiplying $b$ by $10^e$ for an appropriate integer $e$ will move the decimal point over exactly to where the repeating begins, and multiplying $b$ by $10^f$, where $f = e + k$, will move the decimal to where the pattern begins its first repetition. Therefore, there are nonnegative integers $e < f$ such that

$$10^e b - 10^f b = s$$

is an integer. Hence, $b = s/(10^e - 10^f)$ is a rational number.

---

[3] However, the condition that the $a_n$ tend to zero as $n$ goes to infinity is not sufficient for convergence. The most famous example of this is the "harmonic series" $\frac{1}{2} + \frac{1}{3} + \frac{1}{4} + \frac{1}{5} + \cdots$ whose partial sums grow larger than any prescribed number; $\frac{1}{2} + \frac{1}{3} + \frac{1}{4} + \frac{1}{5} + \frac{1}{6} + \frac{1}{7} + \frac{1}{8} + \cdots > \frac{1}{2} + (\frac{1}{4} + \frac{1}{4}) + (\frac{1}{8} + \frac{1}{8} + \frac{1}{8} + \frac{1}{8}) + \cdots = \frac{1}{2} + \frac{1}{2} + \frac{1}{2} + \cdots$, which clearly grows without bound.

For example, let

$$x = \qquad 1.153232323\ldots.$$

Then

$$100x = \qquad 115.323232323\ldots \qquad (7.9)$$

and

$$10000x = 11532.323232323\ldots. \qquad (7.10)$$

When we subtract (7.9) from (7.10), the tails of the two decimals cancel, and we have

$$10000x - 100x = 9900x = 11532 - 115 = 11417$$

and

$$x = \frac{11417}{9900}.$$

Conversely, given the rational number $\frac{s}{t}$, just do long division of $t$ into $s$. There are only $t - 1$ possible remainders at any step, so eventually the remainders must begin to repeat. This means that the digits in the quotient are also repeating, and you get an eventually repeating decimal for your answer.

A second example of a convergent infinite series is the infinite geometric series with ratio $z$:

$$1 + z + z^2 + z^3 + \cdots.$$

If $|z| < 1$, this series converges to the limit $\frac{1}{1-z}$.

These two examples coincide in the case of an infinite decimal such as $1.11111111\ldots.$ This is the infinite geometric series starting with 1 and with ratio $\frac{1}{10}$: $1.11111111\ldots = 1 + \frac{1}{10} + \frac{1}{10^2} + \frac{1}{10^3} + \cdots.$ The formula for the sum of an infinite geometric series in this case gives $\frac{1}{1-1/10} = \frac{1}{9/10} = \frac{10}{9}.$

## 7. $e$, $e^x$, and $e^z$

Infinite series are used to define the number $e$ and the exponential function $e^x$. Let's do this now. The number $e$ is an irrational real number approximately equal to 2.718281828. It is defined to be the limit of the infinite series

$$1 + \frac{1}{1!} + \frac{1}{2!} + \frac{1}{3!} + \frac{1}{4!} + \cdots .$$

This number comes up in elementary calculus and is also present implicitly in trigonometry. It comes up in calculus because of the special property of the function $e^x$.

Recall one way to define $e^x$, for any real number $x$. We start with integers:

- $e^0 = 1$.
- $e^n = \overbrace{e \cdot e \cdots e}^{n \text{ times}}$ if $n$ is a positive integer.
- $e^{-n} = 1/e^n$ if $n$ is a positive integer.

These formulas define $e^n$ for any integer $n$. If $m$ is a positive integer, then we define $e^{1/m}$ to be that positive number whose $m$th power is $e$. If $n \neq 0$ and $m > 0$, set $e^{n/m} = (e^{1/m})^n$. Now we have defined $e^x$ for $x$ any rational number. You can see that $e^x$ is always positive, never negative or zero.

Finally, if $x$ is an irrational number, we can approximate $x$ more and more closely by rational numbers. Choose rational numbers $r_k$ such that $\lim_{k \to \infty} r_k = x$. Then define $e^x = \lim_{k \to \infty} e^{r_k}$. You have to check that this limit exists and that the answer does not depend on the choice of the sequence $r_k$. All of this is verified in an analysis textbook. In a calculus class, you learn that $e^x$ is its own derivative:

$$\frac{d}{dx}(e^x) = e^x.$$

In fact, $e^x$ is the only function $f(x)$ such that $f'(x) = f(x)$ and $f(0) = 1$. It is the differential property of $e^x$ that makes $e$ such an important number, and the function $e^x$ (called the "exponential function") such an important function.

One could write a whole book about $e$, such as Maor (2009). The inverse function of $e^x$ is called the *natural logarithm* function, which we will always denote by $\log x$.

By the way, in our definition of $e^x$, where we started with integral exponents and got up to all real exponents, we could have worked with any positive number $a$ in place of $e$. (Negative $a$ wouldn't work because negative numbers don't have square roots within the context of real number calculus.) We can define the function $a^x$ in a similar way, and it almost satisfies the same differential equation as the one for $e^x$ earlier. But there is a constant that comes in:

$$\frac{d}{dx}(a^x) = ca^x.$$

This constant $c$ depends on $a$. In fact, $c = \log a$.

We need to define $e^z$, where $z = x + iy$ is any complex number. Then we can let $e^z$ be a function on **C**. The best way to define $e^{x+iy}$ is to prove in calculus that $e^x$ can be represented by its convergent Taylor series around $x = 0$:

$$e^x = 1 + \frac{x}{1!} + \frac{x^2}{2!} + \frac{x^3}{3!} + \frac{x^4}{4!} + \cdots.$$

If you remember Taylor series, this is easy to derive from the differential equation for $e^x$ earlier. Every derivative of $e^x$ is just $e^x$ again, and so every derivative has value 1 when $x = 0$. The factorials in the denominator are the standard ones that appear in every Taylor series.

Because of these factorials, this series is convergent for every value of $x$. We could have defined $e^x$ by this series, and we *will* use it to define $e^z$. If you plug any complex number $z$ into the series, you can define

$$e^z = 1 + \frac{z}{1!} + \frac{z^2}{2!} + \frac{z^3}{3!} + \frac{z^4}{4!} + \cdots. \qquad (7.11)$$

This means: Substitute any complex number $z$ into the right-hand side. You get a convergent series whose limit is some complex number, and that answer is by definition $e^z$. Hence, raising $e$ to a complex power, which may seem weird at first, is not a problem.

## 8. Power Series

Formula (7.11) is a specific example of a power series, which we will now describe in more generality. If we put coefficients in front of the terms in the geometric series, and if we let the ratio be a variable rather than a fixed number, we get a power series:

$$a_0 + a_1 z + a_2 z^2 + a_3 z^3 + \cdots .$$

Here we can let $a_0, a_1, a_2, \ldots,$ be any complex numbers, and we can view the series as a function of the variable $z$. In even more generality, we can let $c$ be a fixed complex number, and we can study the series

$$a_0 + a_1 (z - c) + a_2 (z - c)^2 + a_3 (z - c)^3 + \cdots .$$

Again, we think of $z$ as a variable complex number.

It is useful at times to think of a power series as a sort of polynomial with infinitely high degree. In fact, many of the properties of polynomials are also true of power series with appropriate modifications.

It's not too hard to prove that a power series has a *radius of convergence R*, meaning:

- If $R = 0$, then the series converges only if $z - c = 0$ or, equivalently, $z = c$.
- If $R$ is a positive number, the series converges for those $z$ with $|z - c| < R$ and does not converge for those $z$ with $|z - c| > R$.
- We also allow $R$ to be $\infty$, which is a brief way to say that the series converges for all complex numbers $z$.

We omit any statement about what happens for those $z$ for which $|z - c| = R$. In fact, the answer depends crucially on the values of $z$ and the coefficients $a_i$, and can be very hard to determine.

If the radius of convergence $R$ is not 0, then the series defines a function $f(z)$ in a disc of radius $R$ centered at $c$. In this case, the power series is exactly the Taylor series of $f(z)$ about $c$, and we call this disc "the disc of convergence for $f$ about $c$."

An important theorem from complex analysis states:

**THEOREM 7.12:** *Given an open set $\Omega$ in $\mathbf{C}$, and a function $f : \Omega \to \mathbf{C}$, then $f$ is analytic if and only if, for each number $c$ in $\Omega$, the Taylor series of $f(z)$ about $c$ has a positive radius of* *convergence, and converges to $f(z)$ wherever the disc of convergence overlaps with $\Omega$.*

This theorem has two important consequences. First, any function defined by a convergent power series with positive radius of convergence is automatically analytic. Second, any analytic function $f(z)$ automatically has higher derivatives $f^{(k)}(z)$ for all positive integers $k$.

Power series give us a flexible way of studying functions, at least analytic functions, and a lot of great functions are analytic. We've just seen that $\frac{1}{1-z}$ is analytic in the open unit disc $\Delta^0$—or, in other words, the geometric series has radius of convergence 1. Another example of an analytic function is the exponential function $e^z$. We saw its Taylor series in (7.11):

$$e^z = 1 + z + \frac{1}{2!}z^2 + \cdots + \frac{1}{n!}z^n + \cdots .$$

This series has infinite radius of convergence.

We can differentiate and integrate power series, and the two resulting power series will have the same radius of convergence as the one we started with. For example, taking the integral between $z = 0$ and $z = w$ of

$$\frac{1}{1-z} = 1 + z + z^2 + z^3 + \cdots$$

gives

$$-\log(1 - w) = w + \frac{1}{2}w^2 + \frac{1}{3}w^3 + \frac{1}{4}w^4 + \cdots .$$

This equation defines a branch of the logarithm as an analytic function on the open unit disc centered at 1. (See page 122 for a bit more about the concept of a branch of the logarithm function.)

Another flexibility we have with power series is substituting a *function* for $z$. Suppose we have an analytic function

$$f(z) = a_0 + a_1 z + a_2 z^2 + a_3 z^3 + \cdots$$

with radius of convergence $R$. Suppose $w$ is a variable that ranges in some open set $S$, and suppose that $g$ is a complex-valued function on $S$ with the property that $|g(w)| < R$ for every $w$ in $S$. Then we can substitute $g(w)$ for $z$ and get a new function of $w$:

$$f(g(w)) = a_0 + a_1g(w) + a_2g(w)^2 + a_3g(w)^3 + \cdots.$$

This new function is called "the composite of $f$ and $g$." If $g(w)$ is analytic, then so is $f(g(w))$. The chain rule from calculus carries over and tells you how to compute the complex derivative of $f(g(w))$ if you know $f'(z)$ and $g'(w)$, namely

$$\frac{d}{dw}f(g(w)) = f'(g(w))g'(w).$$

You can imagine writing the power series for $g(w)$ wherever you see $g(w)$ and squaring it, cubing it, and so on, and combining all the terms with a given exponent of $w$ and getting the Taylor series of $f(g(w))$. You can imagine it, but you wouldn't want to spend your life doing it—you wouldn't get very far. Sometimes, though, you can determine what all the coefficients of the power series in $w$ have to be without doing all that work.

For example, if $f(z) = z^3$ and $g(w) = (1+w)^{1/3}$, then $f(g(w)) = 1 + w$. Let's partially verify this the hard way. We saw in (7.7) that

$$(1+w)^{1/3} = 1 + \frac{1}{3}w - \frac{1}{9}w^2 + \frac{5}{81}w^3 - \frac{10}{243}w^4 + \cdots. \qquad (7.13)$$

Cubing successively more terms on the right-hand side of (7.13) gives polynomials that are increasingly closer to $1 + w$:

$$\left(1 + \frac{1}{3}w\right)^3 = 1 + w + \frac{1}{3}w^2 + \cdots,$$

$$\left(1 + \frac{1}{3}w - \frac{1}{9}w^2\right)^3 = 1 + w - \frac{5}{27}w^3 + \cdots,$$

$$\left(1 + \frac{1}{3}w - \frac{1}{9}w^2 + \frac{5}{81}w^3\right)^3 = 1 + w + \frac{10}{81}w^4 - \cdots,$$

$$\left(1 + \frac{1}{3}w - \frac{1}{9}w^2 + \frac{5}{81}w^3 - \frac{10}{243}w^4\right)^3 = 1 + w - \frac{22}{243}w^5 + \cdots.$$

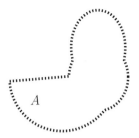

Figure 7.2. An open set

For another example, if $f(z) = e^z$ and $g(w) = -\log(1 - w)$, and you substitute the series for $g$ into the series for $f$, you will get the series $1 + w + w^2 + w^3 + \dots$. That is to say, you just get the geometric series. To say it differently,

$$e^{-\log(1-w)} = \frac{1}{1-w}.$$

This shows that the exponential and the logarithm are inverse functions of each other, something we already know from calculus. If you are so inclined, you could try to substitute one power series into the other and check our assertion by collecting the first few terms.

## 9. Analytic Continuation

One of the most important uses of power series is to define the analytic continuation of analytic functions. For example, the logarithm can actually be defined as an analytic function in much larger domains than just the open unit disc around 1. We discussed this topic in Ash and Gross (2012, pp. 192–95). Here is a brief review.

Discs are not the be-all and end-all of shapes in the complex plane $\mathbf{C}$. You could have an open set $A$ in the plane that looks like figure 7.2, and you could have a function $f : A \to \mathbf{C}$. We know from theorem 7.12 that $f$ is analytic if and only if every point $a$ in $A$ possesses an open disc $D$ centered at $a$ and contained in $A$ with the

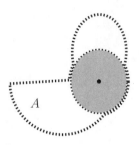

Figure 7.3. The largest open disc centered at $A$

property that $f$ is represented in $D$ by a convergent power series (i.e., by its Taylor series).

Now suppose we have the function $f : A \to \mathbf{C}$, and suppose it is analytic. What can we say about its Taylor series at various points in $A$? How do the different series relate to each other? For a specific function $f$, such questions about the Taylor series can be very hard to answer. However, we can at least imagine a theoretical way of describing what's going on, which we do in the next paragraph.

If $a$ is a point of $A$, then there will be the largest open disc $U$ centered at $a$ and contained in $A$ (see figure 7.3). Theorem 7.12 tells us that the Taylor series of $f$ at $a$ actually converges in all of $U$. Will it perhaps converge in a somewhat larger open disc centered at $a$? We are asking: Is the radius of convergence of this Taylor series larger than the radius of $U$? Maybe it is. If so, we can increase $A$, making it bigger by adding this larger open disc to it. We get a larger domain $A'$ for the function $f$. We can keep doing this, by looking at another point $a'$ in $A'$, getting larger and larger domains $A'', A'''$, and so on, to which we can extend the function $f$, keeping it analytic on the whole larger domain.

In this way, we can get a maximal domain $M$ on which $f$ can be defined. There is a problem that comes up that we will merely mention. There may not be a single natural maximal domain, because you might go around in a circle and when you get back to where you started from, the value of $f$ may not equal what it was at the beginning. This phenomenon is called *monodromy*. To deal with it, you have to leave the world of the plane and start pasting discs together in some abstract world. Then you do get a natural maximal domain, called the *Riemann surface* of $f$.

The process of extending $f$ to its maximal domain is called "analytic continuation." The main point *is that the values of $f$ everywhere in M are determined by the Taylor series centered at a, with which we started.*

One good example is the $\Gamma$-function, which we will encounter again in a later chapter. For a complex number $z$ for which $\text{Re}(z)$ is sufficiently large, we define $\Gamma(z)$ with the equation

$$\Gamma(z) = \int_0^\infty e^{-t} t^{z-1} \, dt.$$

The restriction that $\text{Re}(z)$ is large ensures that the integral converges.

It is not hard to check using integration by parts that

$$\Gamma(z+1) = z\Gamma(z).$$

If $z$ is a bit to the left of where we started, we can then *define* $\Gamma(z)$ with the equation $\Gamma(z) = \Gamma(z+1)/z$. Repeated use of this formula lets us compute $\Gamma(z)$ for any value of $z$ except for $z = 0, -1, -2, -3, \ldots$. (To define $\Gamma(0)$, we would have to divide by 0. And once $\Gamma(0)$ is undefined, we have no way to define $\Gamma(n)$ if $n$ is any negative integer.) We have analytically continued the $\Gamma$-function to all of the complex plane except for the nonpositive integers.

Here is another example. We use the definition of $e^z$ to define $n^s$, where $n$ is a positive real number and $s$ is a complex variable. (For some reason, it is traditional to use the letter "$s$" in this context.) Namely:

**DEFINITION**: $n^s = e^{s \log n}$. Remember that $\log n$ is the logarithm of $n$ to the base $e$.

We mentioned before that the chain rule is true for analytic functions. That is to say, the composition of analytic functions is analytic, and you can use the ordinary chain rule to find the complex derivatives of the composite function. Therefore, $n^s$ is an *entire function* (i.e., analytic for all values of $s$). We form the sum

$$Z_k(s) = \frac{1}{1^s} + \frac{1}{2^s} + \frac{1}{3^s} + \cdots + \frac{1}{k^s}.$$

(Of course, $\frac{1}{1^s}$ is just the constant function 1, but we wrote it that way to enforce the pattern.)

Because $Z_k(s)$ is a finite sum of analytic functions, it is analytic. Now it is not always true that an *infinite* sum of analytic functions is analytic. For one thing, if the infinite sum doesn't converge, it's not clear that it would define a function at all. However, it happens that if you take any open set $A$ on which the real part of $s$ stays greater than 1 no matter what the imaginary part of $s$ is doing,[4] then the limit of $Z_k(s)$ as $k \to \infty$ exists for each $s$ in $A$ and does define an analytic function on $A$. We call this limit function $\zeta(s)$, the Riemann $\zeta$-function:

$$\zeta(s) = \frac{1}{1^s} + \frac{1}{2^s} + \frac{1}{3^s} + \cdots .$$

Okay, we haven't done any analytic continuation yet. It turns out that we can, and that the maximal domain of $\zeta(s)$ is $A = \mathbf{C} - \{1\}$, the whole complex plane except for the number 1. The Riemann hypothesis (RH) says that any value $s = x + iy$ for which $\zeta(x + iy) = 0$ (we're talking about the extended function here) satisfies *either* $y = 0$ and $x$ is a negative even integer *or* $x = \frac{1}{2}$. (A standard reference is Titchmarsh (1986).) No one has proved the RH yet, and it is probably the most famous unsolved problem in mathematics these days.

What makes the Riemann hypothesis so hard to prove or disprove is the mysterious nature of analytic continuation. The function $\zeta(s)$ is completely determined for all $s \neq 0$ by its values in a small open disc about $s = 2$ (for example), but *how* exactly this determination works is opaque.

For our last set of examples in this chapter, we will look at a variant of the $\zeta$-function. We went from a geometric series to a

---

[4] For example, an open disc of radius $r$ centered at $r + 1$, or an open strip where the real part of $s$ lies between 1 and $B > 1$, or maximally the "right half $s$-plane" defined by $\mathrm{Re}(s) > 1$.

power series by inserting coefficients in front of each term. We can do this with the zeta function and define the series

$$a_1 \frac{1}{1^s} + a_2 \frac{1}{2^s} + a_3 \frac{1}{3^s} + \cdots,$$

where the $a_i$ are complex numbers. This is called a *Dirichlet series*. If the norms of the $a_i$ aren't too big, this series will converge to be an analytic function with the variable $s$ restricted to some right half-plane $A$. If the coefficients are "coherent" in some way, the function so defined on $A$ will have an analytic continuation to a larger domain, sometimes all of $\mathbf{C}$. This mysterious coherence often occurs when the coefficients come out of some number theory problem. We will see examples of this in chapter 16, section 3.

*Chapter 8*

# CAST OF CHARACTERS

In this chapter, we introduce or review some crucial numbers, sets of numbers, and functions for use in later chapters. We will first discuss an important subset of the complex numbers and yet more facts about $e^z$. We won't really need $q$ until we get to modular forms, but its definition fits nicely into this chapter.

## 1. $H$

There are various subsets of $\mathbf{C}$ that play important roles in the study of modular forms. The most important of these is the "upper half-plane," usually denoted by some form of the letter "H." In this book, we will use $H$. Here is the very important definition:

$H = \{$complex numbers whose imaginary part is positive$\}$.

It is shown shaded in figure 8.1. Can you see why $H$ is an open set?

The upper half-plane $H$ is also a natural model for non-Euclidean geometry, discovered in the nineteenth century, in which role it can be called the *hyperbolic plane*. In that geometry, the non-Euclidean straight line between two points is the arc of the circle that goes through those two points and makes right angles with the real axis. There is one special case: If the two points have the same real part, then the non-Euclidean straight line between them is the vertical ordinary line that connects them. See figure 11.3. Of course, we only take the parts of the circles and lines that lie in $H$—as far as $H$ is concerned, the rest of $\mathbf{C}$ is "off-limits." Non-Euclidean geometry will be briefly explored in chapter 11.

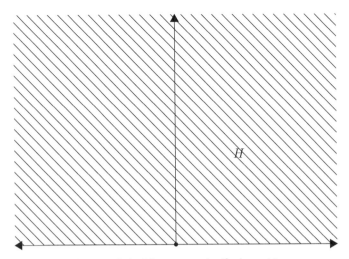

Figure 8.1. The upper half-plane $H$

Rather than let our complex variable $z$ range over all complex numbers, it turns out that we get functions better suited to our purposes if we limit $z$ to take values in $H$.

## 2. $e^z$ Again

We need some more facts about the complex exponential function $e^z$. Using (7.11), the Taylor series for $e^z$, it is not too hard to check some important properties of the exponential function. First of all, if $z$ and $w$ are any two complex numbers, and $n$ is any integer, then

$$e^{z+w} = e^z e^w, \tag{8.1}$$

$$(e^z)^n = e^{nz}.$$

Next, let $z = iy$ be pure imaginary. Then the even powers of $z$ are real, and the odd powers of $z$ are pure imaginary. We can separate them out in the power series on the right-hand side of our defining equation for $e^z$ and can recognize the power series for the cosine and sine functions, as given by their respective Taylor series.

This proves that for any real number $y$

$$e^{iy} = \cos(y) + i\sin(y). \tag{8.2}$$

This is a beautiful formula.[1] If we set $y = \pi$ or $2\pi$, we get the famous formulas

$$e^{\pi i} = -1,$$
$$e^{2\pi i} = 1.$$

We will use these in the next section. Also, notice that because the Pythagorean theorem implies that $\cos^2(y) + \sin^2(y) = 1$, we know that the norm $|e^{iy}| = 1$.

Combining (8.1) and (8.2), we obtain

$$e^{x+iy} = e^x e^{iy} = e^x(\cos(y) + i\sin(y)).$$

This is why we said that the exponential function was lurking behind trigonometry.

You can now see, by plotting $e^{x+iy}$ in polar coordinates, that the norm $|e^{x+iy}| = |e^x||e^{iy}| = |e^x||(\cos(y) + i\sin(y))| = e^x$. And the argument of $e^{x+iy}$ is $y$. So we could have *defined* $e^z$ for any complex number $z = x + iy$ as the complex number with norm $e^x$ and argument $y$. Then we could have worked backward to get the expression for $e^z$ as a power series in $z$.

## 3. $q$, $\Delta^*$, and $\Delta^0$

In the theory of modular forms, it is traditional to use the variable $q$ as

$$q = e^{2\pi i z},$$

where $z$ is a variable ranging over the upper half-plane $H$. This restriction means that $z = x + iy$ where $y > 0$. So $iz = ix - y$, and

$$q = e^{-2\pi y}e^{2\pi i x}.$$

---

[1] We are not the only ones with this opinion; see Nahin (2011).

You can see that the norm of $q$ is $e^{-2\pi y}$ and the argument of $q$ is $2\pi x$. As $x$ moves from left to right along the real axis, $q$ goes around and around in a circular motion, counterclockwise. As $y$ moves from small positive values to large positive values, $q$ gets smaller and smaller in norm. If $y$ is nearly 0, the norm of $q$ is nearly equal to 1, and if $y$ is very large, the norm of $q$ is nearly 0 (but always positive). Putting these facts together, you see that as $z$ roams around the upper half-plane, $q$ fills up the disc of radius 1 centered at the origin 0, *except* $q$ can never be 0. We call this the *punctured disc* $\Delta^*$:

$$\Delta^* = \{w \in \mathbf{C} \mid 0 < |w| < 1\}.$$

It's also useful to fill in the puncture and define the "unit disc" $\Delta^0$ as we did before:

$$\Delta^0 = \{w \in \mathbf{C} \mid 0 \leq |w| < 1\}.$$

So $\Delta^*$ is $\Delta^0$ with 0 removed.

Now why would we want to define $q$? First, notice that $q$ is really a function of $z$. We should write $q(z)$, but typically we won't, because we want to unclutter our notation. As a function of $z$, $q$ has a crucial property:

$$q(z + 1) = q(z).$$

Let's prove this, although it is implicit in the words "around and around" in the previous paragraph. Notice that $z$ and $z + 1$ have the same imaginary part $y$ but the real part $x$ is increased by 1. Hence

$$q(z + 1) = e^{-2\pi y}e^{2\pi i(x+1)} = e^{-2\pi y}e^{2\pi ix+2\pi i} = e^{-2\pi y}e^{2\pi ix}e^{2\pi i} = q(z)$$

because $e^{2\pi i} = 1$.

Very nice—but so what? We say that "$q$ has period 1." In general, a function $g(z)$ has period 1 if $g(z + 1) = g(z)$ for every $z$.

DIGRESSION: The fact that we are only discussing periods of size 1 is for brevity. A periodic function of period $a$ is a function $f(z)$ such that $f(z + a) = f(z)$ for every $z$. We don't let $a = 0$, because in that case this would become a very uninteresting equation. Now if $f(z)$ has period $a$, then we can define a new function $F(z)$ by the formula $F(z) = f(az)$. Obviously, these two functions are closely connected, and if we understand one of them, then we understand the other. But $F(z)$ has period 1 because $F(z + 1) = f(a(z + 1)) = f(az + a) = f(az) = F(z)$. So we don't lose any theoretical strength by discussing just period 1. The number 1 is very nice and simple, so that's why we're going with it. This explains why we put $2\pi$ into the exponent when defining $q$.

Periodic functions are of huge importance in mathematics, physics, and many other sciences. Many physical processes are periodic, or almost periodic. For instance, the Earth turns according to a function that is periodic with a period of 24 hours. The tides are roughly periodic with a period of approximately 24 hours and 50 minutes. The motion of an electron around a hydrogen nucleus in classical physics is periodic, and this periodicity has its implications for quantum mechanics. The motion of a string playing a note on the violin is periodic. And so on.

The mathematical theory of Fourier series was invented to study periodic functions. It is closely connected to $q$ via sines and cosines. You don't need to know about Fourier series to follow what we have to say about $q$, but if you have studied Fourier series, then what we say next may be more believable to you.

We have seen that $q(z)$ is a function on the upper half-plane $H$ with period 1. Could it be the only such function? No, we can easily think of others. Trivially, the constant functions have period 1 (also any other period you like). More interestingly, if we have any function $G$ at all, then $G(q(z))$ also has period 1.

Now "any function $G$" is rather too broad a class of functions— it is uncontrollable. Really nice functions are the algebraic ones— those we get by addition, subtraction, multiplication, and division. Division could be a problem, because we always have to worry about dividing by 0, but we know that $q$ is never 0. Polynomial functions are the nicest: If $a_0, a_1, a_2, \ldots, a_n$ are arbitrary fixed complex numbers, then the function

$$a_0 + a_1 q + a_2 q^2 + \cdots + a_n q^n$$

is a function on the upper half-plane $H$ with period 1.

But we can take limits also. (Compare with chapter 7, section 8.) If we choose an infinity of coefficients $a_n$ and choose them small enough, we can expect that the infinite series

$$a_0 + a_1 q + a_2 q^2 + \cdots + a_n q^n + \cdots$$

will have a limit for every value of $q$ in the punctured disc $\Delta^*$. Therefore, that infinite sum will define a function on the upper half-plane $H$ with period 1. For future reference, note that this function is even defined if we were to set $q = 0$, although $q(z)$ is never 0 for $z \in H$. So it defines a function on the whole disc $\Delta^0$.

We haven't used division yet. We can divide through by any positive power of $q$. This division may cause the appearance of negative powers of $q$, but we repeat: Negative powers of $q$ are not a problem, because $q(z)$ is never 0. Suppose that $m > 0$. Pick fixed complex numbers $a_{-m} \neq 0$, $a_{-(m-1)}, \ldots, a_{-1}, a_0, a_1, a_2, \ldots, a_n \ldots$, so that the $a_n$ get small fast enough as $n \to \infty$. Then the series

$$a_{-m} q^{-m} + a_{-(m-1)} q^{-(m-1)} + \cdots + a_{-1} q^{-1} + a_0 \qquad (8.3)$$

$$+ a_1 q + a_2 q^2 + \cdots + a_n q^n + \cdots$$

will define a function on the upper half-plane $H$ with period 1.

These sums are similar to power series in positive powers of $q$, except they have some extra terms with negative exponents. A series such as (8.3) is called a "Laurent series with a pole of order $m$ at the origin."

We've written down a lot of functions of period 1 on $H$. Could these be all? Certainly not. These power series have a lot of nice properties, and it is not hard to find crazy functions $G$ such that $G(q(z))$ wouldn't have those properties. (For example, every Laurent series defines a *continuous* function on $H$, but we could choose a noncontinuous function $G$.) One theorem in this context, one we will need later for our study of modular forms, comes from complex analysis. It says:

**THEOREM 8.4**: *Suppose $f(z)$ is a function on the upper half-plane $H$ of period $1$, which is analytic and nicely behaved as $y \to \infty$. Then $f(z)$ is equal to some series of the form*

$$a_0 + a_1 q + a_2 q^2 + \cdots + a_n q^n + \cdots,$$

*where $z = x + iy$, $x$ and $y$ are real, and $q = q(z) = e^{2\pi i z}$.*

This is not the book to prove this theorem, but we do owe you an explanation of "nicely behaved." In this context, it means that $\lim_{y \to \infty} f(z)$ exists no matter what $x$ is doing as $y$ goes off to infinity.

By the way, if $f(z)$ possesses all the properties required in the theorem except being nicely behaved, it still can be written as a kind of series in $q$, but now the series may contain terms with negative exponents on $q$—possibly even infinitely many such terms. We use this briefly in Part III.

## Chapter 9

# ZETA AND BERNOULLI

## 1. A Mysterious Formula

There is a mysterious connection between the function $\zeta(s)$ and the Bernoulli numbers $B_k$. Remember that

$$\zeta(s) = 1 + \frac{1}{2^s} + \frac{1}{3^s} + \frac{1}{4^s} + \cdots,$$

and the Bernoulli numbers $B_k$ can be defined with the formula

$$\frac{t}{e^t - 1} = \sum_{k=0}^{\infty} B_k \frac{t^k}{k!}. \tag{9.1}$$

The connection is the formula

$$\zeta(2k) = (-1)^{k+1} \frac{\pi^{2k} 2^{2k-1} B_{2k}}{(2k)!}, \tag{9.2}$$

valid for $k = 1, 2, 3, \ldots$. Note that because $\zeta(2k)$ is positive, (9.2) tells us that the sign of $B_{2k}$ alternates as $k$ increases.

Formula (9.2) is very surprising. Our original reason for defining the numbers $B_k$ (and the related polynomials $B_k(x)$) is that they allowed us to evaluate the *finite* sum $1^k + 2^k + \cdots + n^k$. And now, out of the blue, we can use the same numbers to evaluate an *infinite* sum of the *reciprocals* of even powers.

There are many ways to prove (9.2). The left-hand side can be expressed in terms of double integrals, or in terms of other sums. There are approaches involving Fourier series, and some that use complex numbers. One particularly clever proof (Williams, 1953) establishes the identity

$$\zeta(2)\zeta(2n-2) + \zeta(4)\zeta(2n-4) + \cdots + \zeta(2n-4)\zeta(4) + \zeta(2n-2)\zeta(2)$$

$$= (n + \tfrac{1}{2})\zeta(2n), \tag{9.3}$$

valid for $n = 2, 3, 4, \ldots$. Once $\zeta(2)$ is computed (which is done in Williams (1953) using a formula similar to (9.3)), the values of $\zeta(4)$, $\zeta(6)$, $\zeta(8)$, and so on, can be computed using (9.3) and an identity involving Bernoulli numbers.

We choose instead to follow an approach in Koblitz (1984) that is closer in spirit to Euler's original proof of (9.2). We will omit some technical details in the proof.

## 2. An Infinite Product

**DEFINITION**: A *monic polynomial* in $x$ is a polynomial whose highest power term in $x$ has coefficient 1:

$$x^n + a_1 x^{n-1} + a_2 x^{n-2} + \cdots + a_{n-1} x + a_n.$$

If $p(x)$ is a monic polynomial of degree $n$ with complex coefficients, and $p(\alpha_1) = p(\alpha_2) = \cdots = p(\alpha_n) = 0$ for $n$ distinct complex numbers $\alpha_1, \ldots, \alpha_n$, then $p(x)$ factors as $(x - \alpha_1)(x - \alpha_2) \cdots (x - \alpha_n)$. Notice that the constant term of $p(x)$ is $(-1)^n \alpha_1 \alpha_2 \cdots \alpha_n$.

Suppose on the other hand that $q(x)$ is a polynomial with the property that $q(0) = 1$, and $q(\alpha_1) = q(\alpha_2) = q(\alpha_3) = \cdots = q(\alpha_n) = 0$. In other words, while $p(x)$ was a monic polynomial (with leading coefficient 1), $q(x)$ has *constant* term equal to 1. Take the factorization of $p(x)$ and divide by $(-1)^n \alpha_1 \alpha_2 \alpha_3 \cdots \alpha_n$, and we can write

$$q(x) = \left(1 - \frac{x}{\alpha_1}\right)\left(1 - \frac{x}{\alpha_2}\right)\left(1 - \frac{x}{\alpha_3}\right) \cdots \left(1 - \frac{x}{\alpha_n}\right).$$

Euler made a daring extrapolation. The function $f(x) = \frac{\sin x}{x}$ satisfies $f(0) = 1$ (this assertion is a standard fact from first-semester calculus), and $f(\pm\pi) = f(\pm 2\pi) = f(\pm 3\pi) = \cdots = 0$. Of course, $f(x)$

is 0 for *infinitely* many values of $x$, but Euler nevertheless wrote

$$\frac{\sin x}{x} = \left(1 - \frac{x}{\pi}\right)\left(1 - \frac{x}{-\pi}\right)\left(1 - \frac{x}{2\pi}\right)\left(1 - \frac{x}{-2\pi}\right)$$

$$\times \left(1 - \frac{x}{3\pi}\right)\left(1 - \frac{x}{-3\pi}\right)\cdots .$$

More precisely, he combined pairs of terms and wrote

$$\frac{\sin x}{x} = \left(1 - \frac{x^2}{\pi^2}\right)\left(1 - \frac{x^2}{4\pi^2}\right)\left(1 - \frac{x^2}{9\pi^2}\right)\left(1 - \frac{x^2}{16\pi^2}\right)\cdots . \quad (9.4)$$

This formula is in fact true, and deriving it in a rigorous way is a standard part of a complex analysis course. We will use it without further proof.

We can instantly use (9.4) to compute $\zeta(2)$, and this is in fact what Euler did as soon as he wrote down that equation. If you multiply out the right-hand side, you see that the coefficient of $x^2$ is $-\frac{1}{\pi^2} - \frac{1}{4\pi^2} - \frac{1}{9\pi^2} - \frac{1}{16\pi^2} - \cdots$. How do we handle the left-hand side? We remember from calculus that

$$\sin x = x - \frac{x^3}{6} + \frac{x^5}{120} - \frac{x^7}{5040} + \cdots .$$

Divide by $x$, and we get

$$\frac{\sin x}{x} = 1 - \frac{x^2}{6} + \frac{x^4}{120} - \frac{x^6}{5040} + \cdots .$$

We see that in this expression, the coefficient of $x^2$ is $-\frac{1}{6}$. Therefore,

$$-\frac{1}{6} = -\frac{1}{\pi^2} - \frac{1}{4\pi^2} - \frac{1}{9\pi^2} - \frac{1}{16\pi^2} - \cdots$$

and multiplication by $-\pi^2$ yields

$$\frac{\pi^2}{6} = \frac{1}{1} + \frac{1}{4} + \frac{1}{9} + \frac{1}{16} + \cdots .$$

The right-hand side is $\zeta(2)$, so we conclude that $\zeta(2) = \frac{\pi^2}{6}$.

With some effort, you can study the coefficients of $x^4$ on both sides of (9.4) and compute that $\zeta(4) = \frac{\pi^4}{90}$, and yet more work will show that $\zeta(6) = \frac{\pi^6}{945}$. But a certain amount of cleverness, involving more calculus, gives a better way to derive the connection between $\zeta(2n)$ and $B_{2n}$.

## 3. Logarithmic Differentiation

To make further progress in evaluating $\zeta(s)$ at positive even integers, we substitute $x = \pi y$ in (9.4), cancel some powers of $\pi$, and cross multiply to get

$$\sin \pi y = (\pi y) \left(1 - \frac{y^2}{1}\right) \left(1 - \frac{y^2}{4}\right) \left(1 - \frac{y^2}{9}\right) \left(1 - \frac{y^2}{16}\right) \cdots . \quad (9.5)$$

Now we finish deriving (9.2) by first taking the logarithm of both sides of (9.5) and then differentiating the result.

Start with the right-hand side. Because the logarithm of a product is the sum of the logarithms, the logarithm of the right-hand side of (9.5) is

$$\log \pi + \log y + \sum_{k=1}^{\infty} \log \left(1 - \frac{y^2}{k^2}\right).$$

We can now proceed further. A standard result in calculus is the infinite series

$$\log(1 - t) = -t - \frac{t^2}{2} - \frac{t^3}{3} - \frac{t^4}{4} - \cdots$$

$$= -\sum_{n=1}^{\infty} \frac{t^n}{n},$$

valid for $|t| < 1$. (We saw this formula earlier in chapter 7, section 8.) We therefore have

$$\log \left(1 - \frac{y^2}{k^2}\right) = -\sum_{n=1}^{\infty} \frac{y^{2n}}{nk^{2n}},$$

and so the logarithm of the right-hand side of (9.5) is

$$\log \pi + \log y - \sum_{k=1}^{\infty} \sum_{n=1}^{\infty} \frac{y^{2n}}{nk^{2n}}.$$

Because, for $0 < y < 1$, this sum is absolutely convergent,[1] we are allowed to interchange the two sums to get

$$\log \pi + \log y - \sum_{n=1}^{\infty} \sum_{k=1}^{\infty} \frac{y^{2n}}{nk^{2n}} = \log \pi + \log y - \sum_{n=1}^{\infty} \frac{y^{2n}}{n} \sum_{k=1}^{\infty} \frac{1}{k^{2n}}$$

$$= \log \pi + \log y - \sum_{n=1}^{\infty} \frac{y^{2n}}{n} \zeta(2n).$$

The logarithm of the left-hand side of (9.5) is just $\log \sin \pi y$, so we have

$$\log \sin \pi y = \log \pi + \log y - \sum_{n=1}^{\infty} \frac{y^{2n}}{n} \zeta(2n).$$

Differentiate, and we get

$$\frac{\pi \cos \pi y}{\sin \pi y} = \frac{1}{y} - \sum_{n=1}^{\infty} 2y^{2n-1} \zeta(2n),$$

or

$$\frac{\pi y \cos \pi y}{\sin \pi y} = 1 - \sum_{n=1}^{\infty} 2y^{2n} \zeta(2n).$$

We make one more substitution. Replace $y$ by $z/2$ to get

$$\frac{\pi(z/2)\cos(\pi z/2)}{\sin(\pi z/2)} = 1 - \sum_{n=1}^{\infty} \frac{z^{2n}}{2^{2n-1}} \zeta(2n). \tag{9.6}$$

Remember Euler's formulas for $e^{i\theta}$ and for $e^{-i\theta}$:

$$e^{i\theta} = \cos \theta + i \sin \theta,$$

$$e^{-i\theta} = \cos \theta - i \sin \theta.$$

---

[1] A series $a_n$ of real or complex numbers is said to be *absolutely convergent* if and only if the sum of absolute values $\sum |a_n|$ is convergent. If a series is absolutely convergent, then it is also convergent, but more than that is true. If you rearrange the terms of an absolutely convergent series in any way that you choose, the result is still convergent, and all of those different arrangements converge to the same limit as the original series. Contrast this with a convergent series that is not absolutely convergent, such as $1 - \frac{1}{2} + \frac{1}{3} - \frac{1}{4} + \cdots$. Such a series may be rearranged so that it converges to a different limit, or even so that it diverges.

Addition and subtraction yield

$$\cos\theta = \frac{e^{i\theta} + e^{-i\theta}}{2},$$

$$\sin\theta = \frac{e^{i\theta} - e^{-i\theta}}{2i},$$

and so

$$\frac{\cos(\pi z/2)}{\sin(\pi z/2)} = i\frac{e^{i\pi z/2} + e^{-i\pi z/2}}{e^{i\pi z/2} - e^{-i\pi z/2}} = i\frac{e^{i\pi z} + 1}{e^{i\pi z} - 1}$$

$$= i\frac{(e^{i\pi z} - 1) + 2}{e^{i\pi z} - 1} = i\left(1 + \frac{2}{e^{i\pi z} - 1}\right) = i + \frac{2i}{e^{i\pi z} - 1}.$$

Put all of these changes into (9.6):

$$\frac{\pi i z}{2} + \frac{\pi i z}{e^{\pi i z} - 1} = 1 - \sum_{n=1}^{\infty} \frac{z^{2n}}{2^{2n-1}}\zeta(2n).$$

Remember the defining relationship for the Bernoulli numbers (9.1), and use it in the form

$$\frac{\pi i z}{e^{\pi i z} - 1} = \sum_{k=0}^{\infty} B_k \frac{(\pi i z)^k}{k!} = 1 - \frac{\pi i z}{2} + \sum_{k=2}^{\infty} B_k \frac{(\pi i z)^k}{k!}.$$

Here we used the fact that $B_0 = 1$ and $B_1 = -\frac{1}{2}$. Some cancellation yields

$$\sum_{k=2}^{\infty} B_k \frac{(\pi i z)^k}{k!} = -\sum_{n=1}^{\infty} \frac{z^{2n}}{2^{2n-1}}\zeta(2n).$$

Now, remember that $B_3 = B_5 = B_7 = \cdots = 0$, substitute $k = 2n$, and remember that $i^{2n} = (-1)^n$. We get

$$\sum_{n=1}^{\infty} B_{2n} \frac{(\pi z)^{2n}}{(2n)!}(-1)^n = -\sum_{n=1}^{\infty} \frac{z^{2n}}{2^{2n-1}}\zeta(2n).$$

Equate the coefficient of $z^{2n}$ on each side of the equation, and get

$$\frac{B_{2n}\pi^{2n}(-1)^n}{(2n)!} = -\frac{\zeta(2n)}{2^{2n-1}}$$

or

$$\zeta(2n) = (-1)^{n-1}\frac{2^{2n-1}\pi^{2n}B_{2n}}{(2n)!}.$$

TABLE 9.1. Sum of Reciprocals of Polygonal Numbers

| $k$ | $P(k,n)$ | $\displaystyle\sum_{n=1}^{\infty} \frac{1}{P(k,n)}$ |
|---|---|---|
| 3 | $\dfrac{n^2+n}{2}$ | 2 |
| 4 | $n^2$ | $\dfrac{\pi^2}{6}$ |
| 5 | $\dfrac{3n^2-n}{2}$ | $3\log 3 - \dfrac{\pi\sqrt{3}}{3}$ |
| 6 | $2n^2-n$ | $2\log 2$ |

## 4. Two More Trails to Follow

There is no formula similar to (9.2) known for the values of $\zeta(n)$ for the odd positive integers $n = 3, 5, 7, \ldots$. There are, however, the following alternating sums:

$$1 - \frac{1}{3} + \frac{1}{5} - \frac{1}{7} + \frac{1}{9} - \cdots = \frac{\pi}{4},$$

$$1 - \frac{1}{3^3} + \frac{1}{5^3} - \frac{1}{7^3} + \frac{1}{9^3} - \cdots = \frac{\pi^3}{32},$$

$$1 - \frac{1}{3^5} + \frac{1}{5^5} - \frac{1}{7^5} + \frac{1}{9^5} - \cdots = \frac{5\pi^5}{1536},$$

$$1 - \frac{1}{3^7} + \frac{1}{5^7} - \frac{1}{7^7} + \frac{1}{9^7} - \cdots = \frac{61\pi^7}{184320},$$

with similar results for all odd exponents.

There is another possible generalization of the formula $\zeta(2) = \pi^2/6$. If we view the sum $\zeta(2)$ as the sum of the reciprocals of squares, then we can ask for the sum of the reciprocals of other polygonal numbers. See Downey et al. (2008) for an elegant elementary solution for polygons with an even number of sides. For simplicity, we use the notation $P(3, n)$ to refer to the $n$th triangular number $\frac{n(n+1)}{2}$, $P(4, n)$ to refer to the $n$th square $n^2$, and so forth. The first few results are in table 9.1.

*Chapter 10*

# COUNT THE WAYS

## 1. Generating Functions

How do I sum thee? Let me count the ways.

Often a yes/no problem in mathematics becomes more interesting when it turns into a counting problem. One reason for this is that counting may introduce more structure, or a finer structure, into the data or the conceptual framework of the problem. Similarly, sometimes the counting problem becomes more interesting if it can be turned into a group theory problem. Sometimes, the yes/no problem has an obvious answer but the counting problem still yields a fascinating and beautiful theory. Another reason to solve counting problems is that we are more naturally interested in a more precise and more quantitative question than the simple binary puzzle that we began with. Of course, only with success in achieving an interesting answer will we be satisfied with the more complex approach.

A key example in this book is the problem of sums of squares. We ask not merely "is this number a sum of two squares?" or "which numbers are sums of two squares?" but "*in how many ways* is this number a sum of two squares?" In the case of four or more squares, we know that every positive number is the sum of squares, but the question "in how many ways" is still a good one to ask.

Another example is sums of first powers. This sounds stupid at first: "Is every number the sum of other numbers?" Of course. Any number $n$ is the sum of $0$ and $n$. We usually extend the concept of "sum" to include the sum of one number, and even the sum of no numbers. The "sum" of $n$ (all by itself) is conventionally taken to be $n$.

What is the sum of the empty set of numbers? If you think of how you add numbers on a calculator, you realize that you first have to clear the accumulator. That is, you zero out the register. The screen of your calculator shows 0. Then you start to add your numbers. If you had no numbers to add, the screen stays 0. So the "sum" of no numbers is conventionally taken to be 0. The same calculator fantasy explains why the sum of just $n$ is $n$: You clear the calculator, getting 0, and you add in $n$. You're finished: The answer is $n$.

Let's return to sums of first powers. Here the good question is not whether but in how many ways a given positive integer $n$ can be written as a sum of positive integers, all necessarily less than or equal to $n$. This is called the problem of *partitions*. The problem has generated a huge amount of number theory, a little of which we will discuss later.

We can set up a very general type of problem as follows. Pick a set of integers $S$, which could be a finite set or an infinite set. For instance, $S$ could be the set $\{1, 2, 3\}$, or the set of all prime numbers, or the set of positive integers, or the set of all square integers including 0.

Starting with $S$, we can define a function $a(n)$, where $n$ is any nonnegative integer.

**QUASI-DEFINITION**: $a(n)$ is equal to the number of ways of writing $n$ as a sum of members of $S$.

This quasi-definition is too vague. Are you allowed to use the same member of $S$ more than once in a given sum? Are you going to count the order of the summands as important? For example, suppose $S$ is the set of all square integers: $S = \{0, 1, 4, 9, \ldots\}$. When finding $a(8)$, do we allow $4 + 4$? For $a(25)$, do $9 + 16$ and $16 + 9$ count as just one way or two ways? Do we count $0 + 0 + 9 + 16$ and $0 + 0 + 9 + 16$ as two different ways (where we have switched the order of the 0's—a rather metaphysical thing to do—but certainly thinkable)?

Which choices to make here depend on the particular problem. The structure we use to attack these problems can depend crucially on the choices we make. The wrong choice may lead to an intractable problem, the right choice to a wonderful theory.

Another way of making the quasi-definition more precise would be to limit the number of members of $S$ you are allowed to use in any given sum. For example, if 0 is a member of $S$ and we make no limitation on the number of summands, then $a(n)$ would always be infinite and not very interesting. For example, when we ask about sums of squares, we specify how many: two squares, four squares, six squares, or whatever.

On the other hand, sometimes we don't want to limit the number of summands. In the partition problem, we ask how many ways $n$ can be written as a sum of positive integers, and we do not need to limit the number to get a nice problem.

Suppose we have specified a set $S$ and a precise definition of the function $a(n)$, so we get a sequence of nonnegative integers

$$a(0), a(1), a(2), a(3), \ldots.$$

We want to determine this sequence, maybe by getting a formula for the function $a(n)$. We might also want to know other properties of this sequence, such as, does $a(n)$ have a limit as $n \to \infty$? Is $a(n)$ equal to 0 infinitely often? Is $a(n)$ ever equal to 0? Does $a(n)$ have cute properties? (Maybe $a(n)$ is a multiple of 5 whenever $n$ is a multiple of 5, for example.) These cute properties may look merely cute, but they can be very beautiful indeed and stimulate some very difficult research in number theory, with far-reaching significance. So what do we do with this sequence to study it? Of course, one thing is just to work on the problem itself. For example, if $S$ is the set of squares, and we let $a(n)$ be the number of ways of writing $n$ as a sum of four squares (after having made some determination as to what we do about possible reordering of summands), then we know from theorem 3.2 that $a(n) \geq 1$ for all $n$.

But a more imaginative thing to do is wrap up all the $a(n)$'s into a function of a new variable. At first blush, this process appears to be artificial and not to get us anywhere, but in fact it gets us very far in many cases. For example, we can write down the power series in $x$ whose coefficients are the $a(n)$'s:

$$f(x) = a(0) + a(1)x + a(2)x^2 + a(3)x^3 + \cdots.$$

We call $f(x)$ the *generating function* for the sequence $a(0)$, $a(1)$, $a(2)$, $a(3)$, .... The amazing thing is that if the sequence $a(n)$ comes from a good number theory problem, the structure of the problem may force certain properties on the function $f(x)$ that enable us to study $f(x)$ and then derive information about the $a(n)$'s. This is one of the main points of the remainder of this book.

Another way to package a sequence is to form a *Dirichlet series*. In this case, we must start the sequence from $n = 1$, not from $n = 0$:

$$L(s) = \frac{a(1)}{1^s} + \frac{a(2)}{2^s} + \frac{a(3)}{3^s} + \cdots .$$

Here we hope for convergence of the infinite series for all $s$ in some right half-plane of complex numbers. That is to say, there should be a positive number $M$ such that the series converges to an analytic function of all complex numbers $s$ whose real part is greater than $M$. Then it defines some function $L(s)$ in that half-plane. We can hope that $L(s)$ will analytically continue to be defined on a larger set of values of $s$. If it does, the values of $L(s)$ at certain points in this larger set are often of great number-theoretical interest.

In the case of a generating function in the shape of a power series or a Dirichlet series, the function may have extra properties that enable us to learn more about it, to work back to get information about the sequence $a(n)$ we built it out of, and ultimately to learn more about whatever number-theoretical system produced the $a(n)$'s.

## 2. Examples of Generating Functions

Even a simple sequence like 1, 1, 1, 1, 1, ... produces an interesting generating function. We have already seen this. For the power series, we get

$$1 + x + x^2 + x^3 + \cdots = \frac{1}{1-x},$$

which defines a function for any $x$ of absolute value less than 1. For the Dirichlet series, we get the Riemann $\zeta$-function:

$$\zeta(s) = \frac{1}{1^s} + \frac{1}{2^s} + \frac{1}{3^s} + \cdots .$$

In both cases, the generating function can be analytically continued to the whole complex plane, except for $x = 1$ in the first case and $s = 1$ in the second. The geometric series we feel we understand very well, but the $\zeta$-function still has many mysteries associated with it. Notably, the Riemann hypothesis is still an open problem (see page 94).

Before we give a couple of examples from the theory of partitions, we should officially define a partition.

**DEFINITION**: A *partition* of the positive integer $n$ is an expression $n = m_1 + m_2 + \cdots + m_k$ where the $m_i$ are positive integers. We count two partitions as the same if the right-hand sides are reorderings of each other. We set $p(n)$ to be the number of partitions of $n$. The numbers $m_1, \ldots, m_k$ are called the *parts* of the partition.

In terms of our general yoga from the preceding section, we are taking the set $S$ to be the set of all positive integers. Then $p(n)$ is the number of ways of writing $n$ as a sum of members of $S$ where we allow repetitions and don't care about the order. Because we don't care about the order, we may as well write the sum in ascending order. For example, the partitions of 4 are 4, $1 + 3$, $2 + 2$, $1 + 1 + 2$, and $1 + 1 + 1 + 1$. There are five of them, so $p(4) = 5$. We make the useful convention that $p(0) = 1$, even though by our definition 0 doesn't have any partitions. It may be surprising, but there is no known simple formula for $p(n)$.

We can write down the power series generating function for $p(n)$:

$$f(x) = p(0) + p(1)x + p(2)x^2 + p(3)x^3 + \cdots .$$

We can vary this scheme. For example, we could consider $p_m(n)$: the number of partitions of $n$ into at most $m$ parts. Or $s(n)$: the number of partitions of $n$ where all the parts are odd. You can see there are endless variations here, and some of them lead to very difficult but interesting mathematics, much of it beyond the scope of this book.

We can take the first step in studying partitions by noticing a different way of finding $f(x)$. This illustrates how we can begin to use the flexibility of the generating function method. We start with the geometric series:

$$\frac{1}{1-x} = 1 + x + x^2 + x^3 + \cdots .$$

Notice—trivially—that the coefficient of $x^n$ is 1. We can interpret this by saying that there is only one partition $n = 1 + 1 + \cdots + 1$ of $n$ with no part exceeding 1. This is a bit weird, but think of the term $x^n$ as telling us this.

Now look at the following geometric series:

$$\frac{1}{1-x^2} = 1 + x^2 + x^4 + x^6 + \cdots .$$

What does the term $x^6$ mean here according to our nascent interpretive scheme? It means that 6 has exactly one partition into parts, all of which equal 2, namely $6 = 2 + 2 + 2$.

This still seems weird, but now let's multiply these two equations together. Don't worry about multiplying infinite series: You can just truncate them to be polynomials of large degree, multiply them as polynomials, and then let the degree tend to infinity. In other words, do what seems natural. You get

$$\left(\frac{1}{1-x}\right) \cdot \left(\frac{1}{1-x^2}\right) = (1 + x + x^2 + x^3 + \cdots)(1 + x^2 + x^4 + x^6 + \cdots).$$

If we start multiplying out the right-hand side and collecting terms of a given exponent in $x$, we get $1 + x + 2x^2 + 2x^3 + 3x^4 + \cdots$. For example, where did the coefficient of $x^4$ come from? There were three ways of getting $x^4$ in the product: $1 \cdot x^4$, $x^2 \cdot x^2$, $x^4 \cdot 1$. These correspond to partitions of 4: $2 + 2$, $1 + 1 + 2$, $1 + 1 + 1 + 1$. How? Consider one of these products $x^d \cdot x^e$. The term $x^d$ is coming from $\frac{1}{1-x}$, so by our rule it must be interpreted as a sum of $d$ 1's. The term $x^e$ is coming from $\frac{1}{1-x^2}$, so by our rule it must be interpreted as a sum of $\frac{e}{2}$ 2's. So the product corresponds to a sum of $d$ 1's and $\frac{e}{2}$ 2's, which add up to $d + e = 4$. In other words, we get a partition of 4 into parts, none of which exceeds 2. There are exactly three of them, and that's the meaning of the coefficient 3 of $x^4$ in this product.

Continuing the pattern, we see that

$$\left(\frac{1}{1-x}\right)\left(\frac{1}{1-x^2}\right)\cdots\left(\frac{1}{1-x^m}\right) = 1 + a_1 x + a_2 x^2 + \cdots,$$

where $a_n$ is the number of partitions of $n$ into parts, none of which exceeds $m$.

This is very cool, and since we have had no qualms about multiplying some infinite series together, we may as well multiply an infinite number of them together. WARNING: Such dealings with infinite series do not always yield valid results—you have to make sure things "converge" in some appropriate sense. In the case at hand, notice that when we throw on the new factor $\frac{1}{1-x^m} = 1 + x^m + x^{2m} + \cdots$, we get no terms in $x$ with exponent less than $m$ (except for the constant term 1). This observation means that we won't disturb any of the coefficients of powers of $x$ with exponents less than $m$, so we can fearlessly multiply these geometric series together for all $m$, and we get the beautiful fact that the generating function for partitions satisfies the equation

$$f(x) = \frac{1}{(1-x)(1-x^2)(1-x^3)\cdots}. \tag{10.1}$$

We remind you that $f(x)$ is the generating function for $p(n)$, the number of partitions of $n$ into parts, with no restriction on the size of the parts. Equation (10.1) was first proved, or noticed, by Euler. We will see in later chapters how the theory of modular forms can be brought to bear on $f(x)$.

By changing the exponents, we can derive other generating functions. For example, the generating function for partitions of $n$ into only odd parts would be

$$f_{\text{odd}}(x) = \frac{1}{(1-x)(1-x^3)(1-x^5)\cdots},$$

and so on.

Here's a very cute theorem we can prove easily with these generating functions (Hardy and Wright, 2008, Theorem 344):

**THEOREM 10.2**: *The number of partitions of $n$ into unequal parts is equal to the number of its partitions into odd parts.*

**PROOF**: We have just seen that

$$f_{\text{odd}}(x) = \frac{1}{(1-x)(1-x^3)(1-x^5)\cdots},  \tag{10.3}$$

and (10.3) can be written very cleverly as a telescoping product:

$$f_{\text{odd}}(x) = \frac{(1-x^2)}{(1-x)} \cdot \frac{(1-x^4)}{(1-x^2)} \cdot \frac{(1-x^6)}{(1-x^3)} \cdot \frac{(1-x^8)}{(1-x^4)} \cdots.$$

Here, on the right-hand side, the exponents on the top are 2, 4, 6, 8, ..., and the exponents on the bottom are 1, 2, 3, 4, ..., so the top factors cancel out half of the bottom factors, leaving only the odd exponents.

Now, we know from elementary algebra that $(a - b)(a + b) = a^2 - b^2$. Set $a = 1$, and rewrite this as $\frac{1-b^2}{1-b} = (1+b)$. If we set $b = x$, we have $\frac{1-x^2}{1-x} = (1+x)$. If we set $b = x^2$, we have $\frac{1-x^4}{1-x^2} = (1+x^2)$. If we set $b = x^3$, we have $\frac{1-x^6}{1-x^3} = (1+x^3)$. And so on. Thus we can rewrite the right-hand side to obtain

$$f_{\text{odd}}(x) = (1+x)(1+x^2)(1+x^3)(1+x^4)\cdots.  \tag{10.4}$$

Now the left-hand side of (10.4) is $1 + \sum_{n=1}^{\infty} p_{\text{odd}}(n)x^n$, where $p_{\text{odd}}(n)$ is the number of partitions of $n$ into odd parts. If we write out the right-hand side as a power series, we get $1 + \sum_{n=1}^{\infty} c(n)x^n$, where $c(n)$ is the number of partitions of $n$ into unequal parts. (Do you see why?) Equating the coefficients of $x^n$ gives our theorem.    $\square$

Let's test the theorem with $n = 5$. There are three partitions of 5 with only odd parts: $5 = 5$, $1 + 1 + 3$, $1 + 1 + 1 + 1 + 1$. There are also three partitions of 5 with unequal parts: $5 = 5$, $4 + 1$, $3 + 2$. Very cute.

Here's another example of a useful generating function. Suppose we want to look at sums of squares. We might take $S$ to be the set of positive square integers. Choose and fix a positive integer $k$, and let $r'_k(n)$ be the number of ways of writing $n$ as a sum of $k$ members of $S$, without regard to the order of the summands, parallel to our definition of partitions. It turns out that if we form the generating power series with coefficients $r'_k(n)$, the resulting function does not seem to be easy to work with.

Instead, we define $r_k(n)$ to be the number of ways of writing $n$ as a sum of $k$ squares, including $0^2$, and—this is important—we count $(-a)^2$ and $a^2$ as *different* ways of writing a square if $a \neq 0$. We also count different orders of adding the squares up as distinct. So $a^2 + b^2$ is one way and $b^2 + a^2$ is a different way, provided that $a \neq b$. You see this is rather different from the way that we counted partitions.

For example, let $k = 2$, and look at sums of two squares. Then $r_2(0) = 1$, because $0 = 0^2 + 0^2$, and there is no other way of writing $0$ as a sum of two squares. However, $r_2(1)$ is bigger than you might have thought: $1 = 0^2 + 1^2 = 0^2 + (-1)^2 = 1^2 + 0^2 = (-1)^2 + 0^2$. So $r_2(1) = 4$.

We form the generating function

$$F_k(x) = r_k(0) + r_k(1)x + r_k(2)x^2 + r_k(3)x^3 + \cdots .$$

We will see later that this function jumps into life when we interpret it as a modular form. This will also give us a clue as to why sums of even numbers of squares (i.e., $k$ even) are much easier to study than sums of odd numbers of squares—which is not to say that either of these problems is all that easy.

When $k = 2$, there is a close relationship between the function $F_2(x)$ and a special kind of complex analytic function called an *elliptic function*. Elliptic functions have lots of nice properties, and Jacobi used them to prove the following theorem:

**THEOREM 10.5**: *If $n > 0$, then $r_2(n) = 4\delta(n)$.*

We better define $\delta(n)$. It is defined by $\delta(n) = d_1(n) - d_3(n)$, where $d_1(n)$ is the number of positive divisors of $n$ that leave a remainder of $1$ when divided by $4$, and $d_3(n)$ is the number of positive divisors of $n$ that leave a remainder of $3$ when divided by $4$. (Remember "Type I" and "Type III" primes? If you do, it should not be surprising that remainders upon division by $4$ would come up here.)

For example, what is $\delta(1)$? Well, $1$ has only one positive divisor, namely $1$, and it leaves a remainder of $1$ when divided by $4$. So $d_1(1) = 1$ and $d_3(1) = 0$, so $\delta(1) = 1 - 0 = 1$. And sure enough, $r_2(1) = 4$.

You might think that we could just divide out by 4 to get a simpler formula, because generically, if $a$ and $b$ are different positive numbers, then $a^2 + b^2 = n$ gets counted eight times, when we remember we have to treat different orders of summands and differently signed square roots as different. However, the counting is more subtle. For example, $8 = 2^2 + 2^2$, and this gets counted only four times (because of $(\pm 2)^2 + (\pm 2)^2$) rather than eight times, because here if we switch the summands, we don't get something new.

We may as well continue to check the formula for $r_2(8)$. There's no other way of writing 8 as a sum of two squares, so $r_2(8) = 4$. Now the positive divisors of 8 are 1, 2, 4, and 8, only one of which leaves a remainder of 1 when divided by 4 and none of which leaves a remainder of 3 when divided by 4. So $\delta(8) = 1$, and the formula works again.

For an example involving 0, look at $r_2(9)$. Since $9 = 0^2 + (\pm 3)^2 = (\pm 3)^2 + 0^2$, we get four ways (not eight, because $0 = -0$). There are no other ways. So $r_2(9) = 4$. The positive divisors of 9 are 1, 3, and 9. Two of these leave a remainder of 1 when divided by 4, and one leaves a remainder of 3 when divided by 4. So $\delta(9) = 2 - 1 = 1$, and the formula works again.

It's quite amusing to work out examples of your own. Note that we can quickly deduce from the formula in the theorem one fairly obvious and one additional fact:

(1) The number of ways of writing $n$ as a sum of two squares (counted as we have stipulated) is always divisible by 4.

(2) $d_1(n) \geq d_3(n)$ for all positive $n$.

Theorem 10.5 can be proved directly, without using generating functions (Hardy and Wright, 2008, Theorem 278). That proof relies on the unique factorization of Gaussian integers (i.e., complex numbers of the form $a + bi$ with $a$ and $b$ ordinary integers). But the generating function approach is more fruitful once we go to sums of more than two squares.

### 3. Last Example of a Generating Function

For our last example of the utility of generating functions, we will manipulate the generating function by differentiating it. Recall the

Riemann $\zeta$-function from before. It is a complex analytic function in $s$, so it can be differentiated with respect to $s$.

Suppose we want to study the set of primes. We might try to use a very simple function:

$$a(n) = \begin{cases} 0 & \text{if n is not prime and} \\ 1 & \text{if n is prime.} \end{cases}$$

Then, adding up $a(n)$ for $n$ running from 1 to $N$ would tell us the number of primes between 1 and $N$ inclusive. Call that number $\pi(N)$. This definition is the first very baby step in studying the Prime Number Theorem, which gives us a good approximation to $\pi(N)$. In its simplest form, the Prime Number Theorem says that

$$\lim_{N \to \infty} \frac{\pi(N)}{N/\log N} = 1.$$

It turns out that $a(n)$ is not a very easy function to work with. Numbers like to be multiplied, so our next guess is to use

$$b(n) = \begin{cases} 0 & \text{if n is not a power of a prime and} \\ 1 & \text{if n is a power of a prime.} \end{cases}$$

This is better but still not so great. The problem is that $b(n)$ does not depend on *which* prime $n$ is a power of, if indeed $n$ is a prime power. Finally, we try

$$\Lambda(n) = \begin{cases} 0 & \text{if n is not a power of a prime and} \\ \log p & \text{if } n = p^m \text{ for some prime} p. \end{cases}$$

The choice of the letter $\Lambda$ is traditional.

We can form the generating Dirichlet function:

$$g(s) = \frac{\Lambda(1)}{1^s} + \frac{\Lambda(2)}{2^s} + \frac{\Lambda(3)}{3^s} + \cdots.$$

Then we have the amazing theorem (Hardy and Wright, 2008, Theorem 294)

$$g(s) = -\frac{\zeta'(s)}{\zeta(s)}. \tag{10.6}$$

This equation hints strongly that properties of $\zeta(s)$ may have powerful consequences for the theory of prime numbers.

Let's prove (10.6). The argument is rather complicated, but it shows you how manipulating the generating functions as functions in various ways, plus continual use of the geometric series formula, can really take you places. The first step is to write down the formula

$$\zeta(s) = \left(\frac{1}{1-2^{-s}}\right)\left(\frac{1}{1-3^{-s}}\right)\left(\frac{1}{1-5^{-s}}\right)\cdots. \qquad (10.7)$$

Equation (10.7) is called the *Euler product* for $\zeta(s)$. Let's see why (10.7) is true. The general term in the product on the right-hand side is $\frac{1}{1-p^{-s}}$, where $p$ is a prime. Using the geometric series formula, we can write

$$\frac{1}{1-p^{-s}} = 1 + p^{-s} + (p^2)^{-s} + (p^3)^{-s} + \cdots.$$

For this formula to be true, we must restrict the "ratio" $p^{-s}$ to be less than 1 in absolute value. We can accomplish this by restricting $s$ to complex values with real part greater than 0. In fact, for later purposes that require better convergence, we require the real part of $s$ to be greater than 1.

Now imagine multiplying all these infinite series together, for all the primes. (Ignore convergence problems—it all works fine when you keep track of the details.) Using the laws of exponents, collect the terms. For example, what is the coefficient of $60^{-s}$ in the product? Well, to get $60^{-s}$ we'd have to multiply $(2^2)^{-s}$, $3^{-s}$, and $5^{-s}$ together. This works because $60 = 2^2 \cdot 3 \cdot 5$ and, by the laws of exponents, the $-s$ in the exponent just goes along for the ride. Note that $(2^2)^{-s}$ comes from the factor $\frac{1}{1-2^{-s}}$, $(3)^{-s}$ comes from the factor $\frac{1}{1-3^{-s}}$, and so on. We get exactly one $60^{-s}$ in this way. And this is the *only* way we can get $60^{-s}$. Why? Because the factorization of integers into a product of primes is *unique*. This is why (10.7) is true. The Euler product actually reformulates, via generating functions, the theorem on unique factorization!

Now the logarithm converts multiplication to addition, so we get

$$\log(\zeta(s)) = \log\left(\frac{1}{1-2^{-s}}\right) + \log\left(\frac{1}{1-3^{-s}}\right) + \log\left(\frac{1}{1-5^{-s}}\right) + \cdots. \qquad (10.8)$$

There is an ambiguity in the logarithm of a complex function. Remember that $e^z = e^{z+2\pi i k}$ for any integer $k$. If $w = e^z$, and conditions are favorable, we can choose for $\log w$ one of the values $z$, $z + 2\pi i$, $z - 2\pi i$, and so on consistently as $z$ varies, so long as $z$ does not vary too much. This is called "choosing a branch" of the logarithm. We can and do choose a branch of $\log(\zeta(s))$ such that (10.8) is valid.

Next, we take the derivatives of both sides with respect to $s$. (Again, ignore convergence issues—the details work out.) Remember that the derivative of $\log(s)$ is $1/s$, and use the chain rule. So, for any analytic function $f(s)$ for which we can define a branch of the logarithm $\log f(s)$, we have $\frac{d}{ds} \log f(s) = f'(s)/f(s)$. We also need to work out $\frac{d}{ds} n^{-s} = \frac{d}{ds} e^{(-\log n)s} = -(\log n)e^{(-\log n)s} = -(\log n)n^{-s}$ using the chain rule and the fact that the exponential function is its own derivative.

On the left-hand side, we get

$$\frac{d}{ds} \log \zeta(s) = \frac{\zeta'(s)}{\zeta(s)}.$$

On the other side, we have, for the term involving the prime $p$,

$$\frac{d}{ds} \log \left( \frac{1}{1-p^{-s}} \right) = -\frac{d}{ds} \log(1 - p^{-s}) = -\frac{(1-p^{-s})'}{1-p^{-s}}$$

$$= \frac{-(\log p)p^{-s}}{1-p^{-s}} = \frac{-\log p}{p^s - 1},$$

where in the last step we multiplied the numerator and denominator by $p^s$.

Putting it all together, we obtain the nice formula

$$\frac{\zeta'(s)}{\zeta(s)} = -\frac{\log 2}{2^s - 1} - \frac{\log 3}{3^s - 1} - \frac{\log 5}{5^s - 1} - \cdots.$$

Now, guess what? Let's use the geometric series formula again:

$$\frac{1}{p^s - 1} = \frac{p^{-s}}{1-p^{-s}} = p^{-s} + (p^2)^{-s} + \cdots = \sum_{m=1}^{\infty} p^{-ms}.$$

Therefore

$$\frac{\zeta'(s)}{\zeta(s)} = -\log(2)\sum_{m=1}^{\infty} 2^{-ms} - \log(3)\sum_{m=1}^{\infty} 3^{-ms} - \log(5)\sum_{m=1}^{\infty} 5^{-ms} - \cdots .$$

The sum of sums is absolutely convergent, so we can rearrange it any way we like. In particular, we can say

$$\frac{\zeta'(s)}{\zeta(s)} = -\sum_{n=1}^{\infty} \Lambda(n)n^{-s} = -g(s).$$

QED, as they say.

If you look at (10.6) thinking only of the definitions of the two generating functions $\zeta(s)$ and $g(s)$, then it has to seem like a miracle. Take the derivative of $\zeta$ and divide by $\zeta$. But $\zeta$ had a very simple definition involving just the positive integers. (Of course, they were raised to a complex power....) What we get is $g$, which records which integers are prime powers and which are not, involving also the logarithms of the primes. We might have preferred to avoid these logarithms, but you can't get too greedy. The logarithms come in because of the raising to the power $s$—they are unavoidable in this formula.

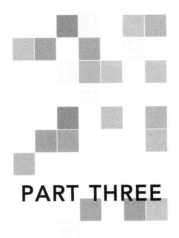

# PART THREE

# Modular Forms and Their Applications

# Chapter 11

# THE UPPER HALF-PLANE

## 1. Review

First, let's review a few concepts from chapter 8. We use $z = x + iy$ to stand for a complex number or variable. So $x$ is its real part and $y$ is its imaginary part. We view $z$ geometrically as a point in the complex plane $\mathbf{C}$, which we identify with the Cartesian $xy$-plane. The absolute value (or norm) of $z$ is $|z| = \sqrt{x^2 + y^2}$. The argument $\arg(z)$ of $z$ is the angle (measured in radians) that a line segment drawn from the origin to $z$ in the complex plane makes with the positive $x$-axis. The upper half-plane $H$ is defined as

$$H = \{\text{complex numbers whose imaginary part is positive}\}.$$

See figures 7.1 and 8.1. Note in addition that $H = \{z \in \mathbf{C} \mid 0 < \arg(z) < \pi\}$.

We will also need to use the function $q(z)$ which we defined as follows:

$$q = e^{2\pi i z}$$

where $z$ is a variable ranging over the upper half-plane $H$. As you can see, we usually omit the dependence on $z$ from the notation in order to make complicated formulas easier to read. We explained that $q(z)$ is a function that sends $H$ onto the punctured unit disc:

$$\Delta^* = \{w \in \mathbf{C} \mid 0 < |w| < 1\}.$$

We will also be using the key property of $q$:

$$q(z + 1) = q(z).$$

From this property, we explained how you can get lots of functions of period 1, and we mentioned the following important theorem:

**THEOREM 11.1**: *Suppose $f(z)$ is a function on the upper half-plane $H$ of period 1, which is analytic and nicely behaved as $y \to \infty$. Then $f(z)$ is equal to some series of the form*

$$a_0 + a_1 q + a_2 q^2 + \cdots + a_n q^n + \cdots .$$

We will now discuss a bit the geometry of the function $q$.

## 2. The Strip

Let's look more closely at the function $q : H \to \Delta^*$. This function is "onto," which means that any complex number $w$ in the punctured unit disc $\Delta^*$ is in the image of $q$. To say it in other words, for any $w \in \Delta^*$, we can always solve the equation

$$q(z) = w$$

for the unknown $z \in H$. This assertion is proved using the logarithm.

Now we'd like to ask, how *many* solutions to this equation are there? Remembering that $q$ is periodic of period 1, we'd have to admit that if $z$ is a solution then so are $z + 1$ and $z - 1$. But repeating this, we see that if $z$ is a solution, then so is $z + k$ for any integer $k$. So we always have an infinite number of solutions.

Now these infinitely many solutions come in a nice package: Their differences yield exactly the set of the integers. An important question is "Are there any other solutions?" To answer this, we have to look more closely at the equation. Using the defining property of the function $q$, we see that we want to solve

$$e^{2\pi i z} = w.$$

A little computation in chapter 8 showed that

$$q = e^{-2\pi y} e^{2\pi i x}.$$

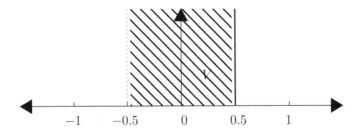

Figure 11.1. The vertical strip $V$

The norm of $q(z)$ is $e^{-2\pi y}$, and the argument of $q(z)$ is $2\pi x$, where, as usual, we are writing $z = x + iy$.

A complex number in the upper half-plane is determined by its norm and its argument. Suppose the norm of the desired $w$ equals $a > 0$ and the argument of $w$ is $\theta$.[1] To find the $z = x + iy$ such that $q(z) = w$ then boils down to finding a value of $y$ such that $e^{-2\pi y} = a$, and a value of $x$ such that $2\pi x - \theta$ is an integral multiple of $2\pi$. There is only one possible $y$, namely $y = -\frac{1}{2\pi}\log(a)$. However, there are infinitely many $x$'s; namely, we can take $x = \frac{1}{2\pi}\theta + k$ for any integer $k$.

Indeed, we had found already that for any value of $z$ solving $q(z) = w$, we get others, namely $q(z + k) = w$ for any integer $k$. Our calculation in the previous paragraph shows us that these are the *only* solutions to $q(z) = w$.

We can express our findings geometrically by considering the "vertical strip" $V$ of all complex numbers $z$ whose real parts lie between $-1/2$ and $1/2$. (Instead of $\pm 1/2$, we could choose here any two real numbers $b$ and $b + 1$ that lie a distance of 1 apart from each other.) To be even more careful, let's specify the boundaries:

$$V = \left\{ z = x + iy \in H \mid -\tfrac{1}{2} < x \leq \tfrac{1}{2} \right\}.$$

See figure 11.1. The left-hand line in the figure is dotted to show that it is *not* part of $V$, and the right-hand side is solid to show that it is part of $V$.

---

[1] Remember that the argument of a complex number $z$ is only determined up to an integer multiple of $2\pi$. Although we usually specify the argument to be in the interval from 0 to $2\pi$, we can always add $2\pi$ to the argument without changing $z$.

Then, a little thought reveals the fact that $q(z) = w$ has one and exactly one solution $z$ in $V$. We call $V$ a *fundamental domain* for the solution set of $q(z) = w$.

For future reference, we can restate this in fancier language. We have a "group" of transformations of $H$ given by shifting the upper half-plane by an integer number of units to the right or the left. In other words, the element called $g_k$ of this group, where $k$ is a particular integer, "acts" on $H$ by the rule

$$g_k(z) = z + k.$$

Then we say that $V$ is a "fundamental domain" for the action of this group on $H$ because any particular point $z \in H$ can be shifted using some $g_k$ so that it becomes a point of $V$, and there is only one shifted point in $V$. To be totally pedantic, we can say that, for any $z \in H$, there exists a unique integer $k$ such that $g_k(z) \in V$. Because of the uniqueness, we can also assert that if $z_1$ and $z_2$ are two points *both in* $V$, and if $g_k(z_2) = z_1$ for some $k$, then in fact $z_1 = z_2$.

We will come back to this terminology in a more complicated case later in this chapter.

### 3. What Is a Geometry?

The upper half-plane is a model for the hyperbolic, non-Euclidean plane. A full explanation of this would nicely fill another volume. We will sketch a little about what is going on here.

Geometry began with measuring plots of land, perhaps in Egypt. (That's what the word means in Greek: "geo-metry" = "land-measuring.") With the ancient Greeks, it became a study of (what appears to be) the space we live in, both two- and three-dimensional. In modern mathematics, the word "geometry" takes on a vastly greater range of meanings.

We may as well stick to "plane geometry"; that is to say, two-dimensional geometry. An example of this is the plane as studied in Euclid's *Elements*. The main objects for Euclid are points, lines, and circles. (In this context, we normally speak of "lines" when we mean "straight lines.") In the *Elements*, you are allowed to connect two

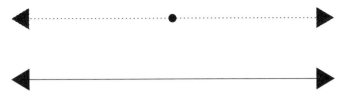

Figure 11.2. The parallel postulate

points with a line segment using a "straightedge" (a ruler without marks). You are also allowed to draw a circle with a compass. You can also transfer a line segment from one place to another using the compass, keeping the length of the segment fixed. Hence we view the Euclidean plane as a "two-dimensional manifold" of points, equipped with a "metric" that measures the distance between any two points. Any two points determine a unique line on which they both lie.

Do any two lines intersect in a unique point? Not quite: Parallel lines do not intersect at all. In Euclidean geometry, given any point and any line that doesn't go through that point, there exists a unique new line that does go through the point and never intersects the first line. See figure 11.2.

The new line and the old line are said to be parallel to each other. Our assertion about the existence of parallel lines is equivalent to the fifth postulate in Euclid's *Elements*, given all the other axioms and postulates. It is called the "parallel postulate."

Suppose you took a book of Euclid's *Elements* but erased all the diagrams. You could then ask for a set of points that instantiates all the axioms and postulates. (Skip the definitions, where Euclid tries to explain what exactly he means by words like "point" and "line.") If you find such a set, you can call it a "model" of Euclidean geometry. For example, the familiar idealized plane that goes on forever in all directions with the usual metric (notion of distance between two points) is a model of Euclidean geometry. The points in the plane are the points, the lines connecting them are the lines, and so on. This is not surprising, because that's the ideal plane that Euclid had in his mind.

But, surprisingly, we can create other models of Euclidean geometry. We wouldn't especially have any reason for the following example, but we could take a paraboloid $P$ in space. The points of $P$ we declare to be our points. Then there is a way of declaring certain curves on $P$ to be our lines, and a way of choosing a metric on $P$ in such a way that all axioms and postulates of Euclidean geometry become true in this setup. And therefore all the theorems of Euclid's *Elements* also become true for it. This is another model of Euclidean geometry, different from the flat plane but equivalent to it as far as all the axioms, postulates, and theorems of Euclidean geometry go.

Another example: We could take a filled-in square $E$ *without its bounding edges* and declare certain sets of points to be lines. We could define a certain metric so that $E$ becomes yet another model of the Euclidean plane. Points near opposite edges of $E$ would be a huge distance apart with respect to this new metric.

Descartes taught us yet another way to create a model of Euclidean geometry. We take the set $D$ of all pairs $(x,y)$ of real numbers. We declare each such pair a point in our model. We declare a line in our model to be any set of points that is the solution set to some equation of the form $ax + by = c$, where $a$, $b$, and $c$ are real constants (and either $a$ or $b$ or both are not zero). We define the metric by saying that the distance between two points $(x,y)$ and $(t,u)$ is equal to $\sqrt{(x-t)^2 + (y-u)^2}$. Again, all axioms, postulates, and theorems of Euclidean geometry become true for $D$ with these declarations. But $D$ contains only pairs of numbers, not any actual geometric points.

## 4. Non-Euclidean Geometry

For centuries, some people tried to prove that the parallel postulate could be derived from the other axioms and postulates. In the nineteenth century, this was shown to be impossible. How? Mathematicians constructed various models in which all the other axioms and postulates held but the parallel postulate failed.

This is something Euclid himself could have done, but the Greeks didn't think that way. Euclid seems to be desiring to describe what he had in his mind—the ideal Euclidean plane. And that's that. We don't know if he or his colleagues attempted to prove the fifth postulate and when they couldn't they just left it as a postulate. It is tempting to conjecture that they did try.

We will now take the upper half-plane $H$ and use it to make a model for a geometry in which all of Euclid's other axioms and postulates hold but the parallel postulate definitely fails. For the points in our model, we take the points of $H$. The lines of our model are a little complicated to describe. Here's how we do it. Any two distinct points $p$ and $q$ in $H$ must determine a unique line. If $p$ and $q$ have the same real part, then we declare the vertical half-line in $H$ connecting them to be a "line" in the model we are constructing. Otherwise, there is a unique semicircle that goes through both $p$ and $q$ and that has its center on the $x$-axis. We declare this semicircle to be a "line" in our model. We can also describe the metric in terms of a certain integral along the "line" that connects $p$ and $q$. Namely, the distance between $p$ and $q$ in the upper half-plane is equal to

$$\int_L \frac{ds}{y},$$

where $L$ is the "line" segment (i.e., the arc of the semicircle or the vertical line segment) that connects $p$ and $q$, and $ds$ is the standard Euclidean distance in $\mathbf{C}$ (i.e., $ds^2 = dx^2 + dy^2$). (If you don't know exactly what this integral means, you can ignore all of these details. You just need to know that there is some exact definition of the distance between two points.)

You can now, if you like, check that all of Euclid's axioms and postulates hold for this model of geometry, except: Given a "line" $L$ through $q$ and a point $p$ not on that "line," there will not be a unique line through $p$ that doesn't intersect $L$; rather, there are infinitely many. In figure 11.3, for example, the two lines $L_1$ and $L_2$ both pass through $p$, and neither line intersects $L$.

Therefore, many of the theorems of Euclid's *Elements* do *not* hold for $H$ with these definitions of point, line, and distance. For example, the angles in a triangle always add up to *less* than 180°.

The use of the word "line" has now become ambiguous. If we look at $H$ and say the word, we might mean a Euclidean line, or we might mean a non-Euclidean line (which in the previous

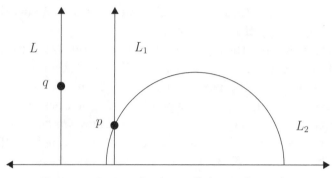

Figure 11.3. Two lines parallel to $L$ through $p$

paragraphs we were putting in quotation marks). To avoid confusion, we often use the word "geodesic" to refer to a line in a non-Euclidean plane.

This model of geometry we have placed on $H$ is called a model of the "hyperbolic plane." There are other models of the hyperbolic plane. For example, we could take instead as our set of points the open unit disc and define geodesics and the metric in some appropriate way, which we need not go into here.

## 5. Groups

Already in the second half of the nineteenth century, a decisive shift in the way many mathematicians look at geometry was undertaken, under the leadership of Felix Klein. The basic idea is helpful in motivating our interest in a certain "group" that will be crucial in our study of modular forms. There couldn't be a theory of modular forms without this group.

First, what is a group?[2] It is a set on which we have defined a "multiplication." This operation need not have anything to do with ordinary multiplication. It is just a specified way of taking two elements of the set and combining them to get a third. We write

---

[2] We will be very brief here. For a lengthier exposition, you could look at Ash and Gross (2006, chapters 2 and 11).

the multiplication of $g$ and $h$ simply by juxtaposing them: $gh$. A group must contain a special element, called the "neutral element," which leaves things alone under multiplication. For instance, if the neutral element is $e$, then $ge = eg = g$ for any $g$ in the group. Notice that we write both $ge$ and $eg$, because there is nothing about a group that says that multiplication has to obey the commutative law.

Another necessary property of a group is that for any element $g$ in the group there has to be some (uniquely determined) element $j$ such that $gj = jg = e$. This $j$ depends on $g$, of course, and is written $g^{-1}$, spoken "$g$-inverse." For example, $ee = e$, so $e$ is its own inverse, and we write $e = e^{-1}$. Usually, the inverse of an element is a different element, but there can be elements other than $e$ that are their own inverses. Or there may not be. It depends on the group.

The final property needed for a group is the associative law of multiplication: $(gh)k = g(hk)$ for any elements $g$, $h$, and $k$ in the group. Because of the associative law, we typically omit the parentheses and just write $ghk$ for the triple product.

There are zillions of groups—in fact infinitely many kinds of groups, and often infinitely many instances of each kind. In this book, the only groups we will need to discuss in detail are groups of matrices. But you already know some groups. For example, the real numbers **R**, in which we declare the "multiplication" to be ordinary addition, is a group. The neutral element is 0, the inverse of $x$ is $-x$, and you can check all our requirements are satisfied.

As you can see, we can get confused with the word "multiplication." Sometimes we will call the rule by which two elements of the group are combined to create their "product" by a different name: we will call this rule "the group law." So in the previous example, we would say "**R** is a group with the group law of addition."

Now you can think up other groups. For example, you could take the set of integers, with the group law being addition. Or the set of nonzero real numbers, with the group law being ordinary multiplication. (In this example, the neutral element is 1.)

These are all numerical examples, and in each of them, the group law has been commutative (i.e., $gh = hg$ for all pairs of elements in the group). Sets of functions can be used to give us lots of interesting groups, and more often than not the group law is not commutative.

For example, let $T$ be a set with more than two elements. Let $G$ be the set of all one-to-one correspondences $f$ from $T$ to itself. We think of $f$ as a function from $T$ to $T$. Given an element $t$ in $T$, we write the element that corresponds to it under the one-to-one correspondence $f$ using the notation $f(t)$.

We define the group law as composition of functions: If $f$ and $g$ are in $G$, then $fg$ is defined to be the one-to-one correspondence that sends $t$ to $f(g(t))$ for any $t$ in $T$. (Note carefully the order of $f$ and $g$ in this definition.) With this law, $G$ is a group. The neutral element $e$ is the one-to-one correspondence that sends each element in $T$ to itself: $e(t) = t$ for all $t$ in $T$. This group is never commutative (because we assumed that $T$ contains at least three elements).

Now we can explain Felix Klein's idea, cut down to fit our context. Klein said to look at a model of some geometry you were interested in. Suppose the model consists of the set $T$ with a given notion of points, geodesics, and distance. Consider the set $G$ of all one-to-one correspondences $f : T \to T$ that *preserve* these notions. In other words, $f$(a point) = a point and $f$(a line) = a line.[3] Moreover, if the metric tells you the distance between the two points $t$ and $u$ is some number $d$, then the distance between the two points $f(t)$ and $f(u)$ is equal to the same number $d$.

Notice that $G$ always has at least one member, namely the neutral one-to-one correspondence $e$. It is easy to check that $G$, with the group law of composition of functions, is a group. This group obviously can have a lot to do with the geometry you started with. If that geometry is sufficiently "homogeneous," then you can actually reconstruct it from the knowledge of the group $G$ alone. This enabled Klein to suggest a general method for creating new geometries, starting from sufficiently "homogeneous" groups given abstractly. His point of view has been extremely productive.

For example, let $T$ be a model for the Euclidean plane. To be concrete, let's take it to be the Cartesian model we called $D$. If we form the group $G$ for this geometry, what kinds of elements does it have? We have to think of one-to-one correspondences from $D$ to

---

[3] We adopt here the usual and useful notation that if $A$ is any subset of $T$, then $f(A)$ denotes the subset consisting of all the images $f(a)$ for all $a$ in $A$.

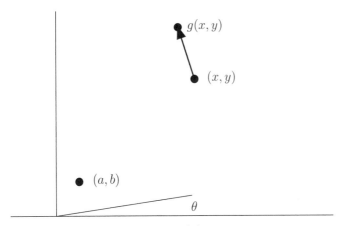

Figure 11.4. A Euclidean motion

itself that preserve points, lines, and distances. Because the points of our geometry are just the points of the set $D$, preserving points is automatic. Because a line is the unique curve of shortest distance between two points (a nontrivial fact), if we preserve distance we will also preserve lines. So we are just asking: Find all the one-to-one correspondences from $D$ to itself that preserve distances.

Well what can you do? You can rotate $D$ about some point, and you can translate $D$ along some vector. You can also compose these two types of functions. It turns out that that's it. The group $G$ consists of all one-to-one correspondences from $D$ to itself that you get by rotating by a certain number of degrees about some fixed point and then translating in a certain direction by a certain fixed distance.

This group $G$ is traditionally called the *group of Euclidean motions*. We can describe an element of $G$ using formulas. It turns out that you can get every element of $G$ by rotating counterclockwise around $(0,0)$ by some number of radians, say $\theta$, and then translating by a given vector, say $(a,b)$. We can write this as the formula

$$(x,y) \rightarrow g(x,y) = (a + x\cos\theta - y\sin\theta, b + x\sin\theta + y\cos\theta).$$

See figure 11.4.

Well, we lied. We only described half of all the elements of $G$. We can, if we wish, follow any motion by a reflection that flips the plane along some given line. For instance, the flip along the $y$-axis is given by the formula

$$(x, y) \to (-x, y).$$

If you like, you can add this into the preceding formula and get a formula for the most general Euclidean motion in $G$.

Klein taught that points, lines, and distances are not sacred. Sometimes you might want to forget about distances, for example. In this way, you can define more general geometries, such as projective geometries, and even finite geometries, where there are only a finite number of points and lines!

In the next section, we will show how to describe Euclidean motions using matrices, and then we will be all set to revisit the hyperbolic plane from Klein's point of view.

## 6. Matrix Groups

In this book, we will only need to talk about 2-by-2 matrices. What is a 2-by-2 matrix? It is an array of numbers that looks like

$$\begin{bmatrix} a & b \\ c & d \end{bmatrix}.$$

Here, $a, b, c,$ and $d$ can be any numbers. If we fix a number system,[4] such as $\mathbf{R}$ or $\mathbf{C}$, from which to choose $a, b, c,$ and $d$, then we will call this set of arrays $M_2(\mathbf{R})$ or $M_2(\mathbf{C})$. If we let $A$ be any number system, then we can write $M_2(A)$ for the set of all 2-by-2 arrays with entries taken from the set $A$. The subscript 2 tells us that we are dealing with 2-by-2 arrays.

---

[4] In every number system we consider, we assume that $1 \neq 0$, that addition and multiplication obey the associative and commutative laws, and that multiplication is distributive over addition.

Matrix multiplication is given by the following rule:

$$\begin{bmatrix} a & b \\ c & d \end{bmatrix}\begin{bmatrix} e & f \\ g & h \end{bmatrix} = \begin{bmatrix} ae+bg & af+bh \\ ce+dg & cf+dh \end{bmatrix}.$$

It's a bit complicated at first sight, but it is the right definition for us. Using this rule, you can see that the matrix

$$I = \begin{bmatrix} 1 & 0 \\ 0 & 1 \end{bmatrix}$$

is the neutral element for matrix multiplication. A painstaking bit of algebra will show you that matrix multiplication is always associative.

So does this mean $M_2(A)$ is a group with group law = matrix multiplication? No. The matrix

$$O = \begin{bmatrix} 0 & 0 \\ 0 & 0 \end{bmatrix}$$

has the property that it multiplies everything to $O$. So it cannot have an inverse matrix $O^{-1}$, which would need to satisfy $OO^{-1} = I$. Because the rules for a group require inverses, this observation means that $M_2(A)$ cannot be a group under matrix multiplication.

In fact, you don't have to be so radical to fail to have an inverse. You can check, for example, that

$$\begin{bmatrix} 1 & 0 \\ 1 & 0 \end{bmatrix}$$

has no inverse under matrix multiplication, and there are many other examples.

We have to bite the bullet and move to a subset of $M_2(A)$ smaller than the whole thing. We define this subset, called $GL_2(A)$, to be the set of all 2-by-2 matrices with entries in $A$ that have inverses under matrix multiplication that are also 2-by-2 matrices with entries in $A$. (The letters GL stand for "general linear group.")

Luckily, it turns out there is a simple way to tell if a 2-by-2 matrix has or doesn't have an inverse under matrix multiplication. We define the "determinant" of a 2-by-2 matrix by the following formula:

$$\det \begin{bmatrix} a & b \\ c & d \end{bmatrix} = ad - bc.$$

For example, $\det(I) = 1 \cdot 1 - 0 \cdot 0 = 1$ and $\det(O) = 0 \cdot 0 - 0 \cdot 0 = 0$. The determinant is "multiplicative," meaning $\det(KL) = (\det K)(\det L)$.

**THEOREM 11.2**: *The matrix K has an inverse under matrix multiplication in* $GL_2(A)$ *if and only if* $\det(K)$ *has a multiplicative inverse in A.*

This theorem is not so hard to prove, but we won't prove it. The easy part we will do; namely, if $\det(K)$ has a multiplicative inverse in $A$, we can write down a formula for the inverse of $K$. That, of course, shows that it has an inverse. Here is the formula, assuming that $ad - bc \neq 0$:

$$\begin{bmatrix} a & b \\ c & d \end{bmatrix}^{-1} = \frac{1}{ad - bc} \begin{bmatrix} d & -b \\ -c & a \end{bmatrix}.$$

A MNEMONIC: You switch the diagonal entries, negate the off-diagonal entries, and divide by the determinant. We can now give an alternative definition of the general linear group:

$GL_2(A) =$

$\{K \in M_2(A)$ such that $\det(K)$ has a multiplicative inverse in $A\}$.

For example, if our number system is $\mathbf{R}$, then every nonzero number has a multiplicative inverse in $\mathbf{R}$, so

$$GL_2(\mathbf{R}) = \{K \in M_2(\mathbf{R}) \mid \det(K) \neq 0\}.$$

The analogous definition works for $GL_2(\mathbf{C})$.

On the other hand, if our number system is the set of integers $\mathbf{Z}$, then only 1 and $-1$ have multiplicative inverses in $\mathbf{Z}$, so

$$GL_2(\mathbf{Z}) = \{K \in M_2(\mathbf{Z}) \mid \det(K) = \pm 1\}.$$

Notice that we have to be careful about which number system we are "in." For example, the matrix

$$K = \begin{bmatrix} 1 & 2 \\ 3 & 4 \end{bmatrix}$$

has determinant $-2$ and inverse

$$\begin{bmatrix} -2 & 1 \\ \frac{3}{2} & -\frac{1}{2} \end{bmatrix}.$$

So $K$ is an element of $\mathrm{GL}_2(\mathbf{R})$ but is not an element of $\mathrm{GL}_2(\mathbf{Z})$.

We can use matrices to list all Euclidean motions of the plane. It is not hard to check that each of the following is a Euclidean motion, and some more effort will prove that this list includes all of the Euclidean motions. Thus, the group of Euclidean motions may be described as the set of pairs $(K, v)$, where $K$ is a matrix in $\mathrm{GL}_2(\mathbf{R})$ of the form

$$\begin{bmatrix} \pm\cos\theta & -\sin\theta \\ \pm\sin\theta & \cos\theta \end{bmatrix}$$

and $v$ is a vector of the form

$$\begin{bmatrix} a \\ b \end{bmatrix}.$$

The pair $(K, v)$ corresponds to the motion of the plane that sends a vector $x$ to the new vector $Kx + v$. The group law is a little complicated, and we will drop this discussion here because we do not need it. Things are simpler to describe for the upper half-plane, as follows.

## 7. The Group of Motions of the Hyperbolic Non-Euclidean Plane

Once we adopt Klein's point of view, there are many different kinds of geometries. Even before Klein, mathematicians realized that there were different non-Euclidean geometries of two dimensions, for example the hyperbolic plane and the sphere. How many different geometries exist depends on the precise definition of the term "geometry." In the rest of this book, we will be concerned with only one type of non-Euclidean geometry, called the *hyperbolic plane*. In this context, it is traditional to use the term "non-Euclidean plane" to refer to the hyperbolic plane, in order to stress that it does not obey Euclid's fifth postulate about parallel lines. We will continue this traditional usage here.

Let's take $G$ to stand for the group of motions of the hyperbolic non-Euclidean plane. For our model, we take the upper half-plane $H$. So we are looking for one-to-one correspondences $f : H \rightarrow H$ with the property that $f$ preserves points (which is automatic), geodesics, and the distance.

Before we do, remember that preserving geodesics means that $f$ takes any given vertical half-line to another (possibly the same) vertical half-line or to a semicircle with center on the $x$-axis. It must also take any such semicircle to another such (possibly the same) semicircle or to a vertical half-line.

There is one obvious kind of function $f$ in $G$: translation by a real number $b$. In other words, if $b$ is any real number, then the function on $H$

$$z \rightarrow z + b$$

takes vertical lines to vertical lines and semicircles to semicircles—it is just sliding everything over by $b$. In fact, this function also preserves the distance, and so such functions are in our group $G$.

What else? Another thing we can do is the flip described using complex conjugation:

$$z \rightarrow -\bar{z}.$$

(If $z = x + iy$, then $\bar{z} = x - iy$ and $-\bar{z} = -x + iy$.) This flip also preserves geodesics and the distance function.

A third, slightly less obvious function is dilation. If $a$ is any positive real number, then it is possible to check that multiplication by $a$ preserves geodesics and the distance. We call this "dilation" by $a$:

$$z \rightarrow az.$$

Notice that we had better make sure $a$ is positive, because if $a$ is negative it will take the upper half-plane to the lower half-plane.

Putting all of these together, we get functions of the type

$$z \to az + b \text{ or } a(-\bar{z}) + b.$$

If this were all, we wouldn't have a very interesting group. For one thing, all of these motions extend to the whole plane $\mathbf{C}$. Also, they take half-lines to half-lines and semicircles to semicircles. They don't mix them up. But there is a fourth kind of non-Euclidean motion that makes things get very interesting.

Consider what is called an "inversion":

$$\iota(z) : z \to -1/z.$$

Notice that we are not in danger of dividing by zero. That is because $z$ is in the upper half-plane, so $z \neq 0$. Let's see what happens to the vertical half-line given by $x = 0$. Then $z = iy$ and $\iota(z) = -1/(iy) = i/y$, because $i^2 = -1$ implies that $1/i = -i$. Well, that's not apparently so very interesting: $\iota$ takes this particular vertical half-line to itself—although it does turn it upside-down, which is a bit surprising.

What does it do to another vertical half-line, say that given by $x = 3$? Then $z = 3 + iy$ and

$$\iota(3 + iy) = -\frac{1}{3 + iy} = -\frac{3 - iy}{(3 + iy)(3 - iy)} = -\frac{3 - iy}{3^2 + y^2}.$$

What curve does $w = -\frac{3-iy}{9+y^2}$ describe as $y$ goes from 0 to $\infty$? Set $\xi = -\frac{3}{9+y^2}$ and $\eta = \frac{y}{9+y^2}$, so that $w = \xi + i\eta$. We have to eliminate $y$ and get an equation relating $\xi$ and $\eta$. For short, let's write $t = 9 + y^2$. Then

$$\xi = -\frac{3}{t}, \quad \xi^2 = \frac{9}{t^2}, \quad \text{and} \quad \eta^2 = \frac{y^2}{t^2} = \frac{t - 9}{t^2}.$$

We see that $t = -\frac{3}{\xi}$, so $\eta^2 = (-\frac{3}{\xi} - 9)(\frac{\xi^2}{9}) = -\frac{\xi}{3} - \xi^2$.

We see that $w$ lies on the circle with equation $\xi^2 + \frac{\xi}{3} + \eta^2 = 0$. Complete the square to rewrite this as

$$\left(\xi + \frac{1}{6}\right)^2 + \eta^2 = \frac{1}{36}.$$

This is the equation of the circle with center at $(-\frac{1}{6}, 0)$ and radius $\frac{1}{6}$. As $y$ goes from 0 to $\infty$, $t$ goes from 9 to $\infty$ and $\xi$ goes from $-\frac{1}{3}$ to 0. It is all fitting together: We're not getting the whole circle, but just the part of it lying in the upper half-plane. Because its center is on the real line, we are getting exactly a semicircle, which hits the real axis at right angles. In other words, $\iota$ takes the vertical geodesic given by $x = 3$ to the geodesic consisting of this semicircle.

You may have skipped the previous calculation, but in any case such calculations will show that $\iota$ takes every geodesic to another (possibly the same) geodesic. In fact, $\iota$ preserves the distance function and is in our group $G$.

How do you get more elements of $G$? Well, if you compose any two elements of $G$, you again get an element of $G$. For instance, if you compose the functions $z \to az + b$ and $\iota$, you get $z \to -1/(az + b)$. If we make all possible compositions, we get a set of functions that turn out to be closed under composition, and it can be proven that this is the entire group $G$. There are no more surprises or weird functions we haven't thought of.

Now, $G$ is a bit awkward because it has flips in it. We don't like flips, because they are not complex-differentiable functions[5] of $z$. So from now on we won't allow flips until we have a really good reason to talk about them. We let $G^0$ stand for the group of all nonflipping motions of the non-Euclidean plane.

If we follow out our computations and compositions as just described, we find that every element of $G^0$ can be described in the form

$$z \to \frac{az + b}{cz + d},$$

where $ad - bc > 0$. This inequality is what is needed to map $H$ to $H$.

---

[5] If you compute the difference quotient of $z \to \bar{z}$, you get $(\overline{z + h} - \bar{z})/h$. The derivative at $z$ is supposed to be the limit of this as the complex number $h$ tends to 0. Let's compute the derivative (if it exists). We get $\lim_{h \to 0} \bar{h}/h$. But if $h$ goes to 0 along the real axis, we get 1, while if $h$ goes to 0 along the imaginary axis, we get $-1$. So the limit does not exist, and the complex derivative does not exist either.

This group $G^0$ contains inside it in embryo all of the non-Euclidean hyperbolic plane. Klein tells us to meditate on this group if we want to understand better the hyperbolic plane in its upper half-plane model $H$. That's what hyperbolic geometers do. What we number theorists will do instead is to look at some "subgroups" of $G^0$ and some amazing number theory that can be developed from them.

The element of the group $G^0$ that we just displayed can be written succinctly as a 2-by-2 matrix:

$$\gamma = \begin{bmatrix} a & b \\ c & d \end{bmatrix}.$$

When we do this, we can write the corresponding non-Euclidean motion as

$$z \rightarrow \begin{bmatrix} a & b \\ c & d \end{bmatrix}(z)$$

or more briefly as $z \rightarrow \gamma(z)$. A function of $z$ of this type is called a "fractional linear transformation."

Now here is an exercise for you to do that will test how well you followed all of the preceding discussion. The group law of Euclidean motions is composition of functions. The group law of matrices is matrix multiplication. The great thing is that these two group laws are compatible. If $\gamma$ and $\delta$ are two matrices in $M_2(\mathbf{R})$ with positive determinants, then

$$\gamma(\delta(z)) = (\gamma\delta)(z).$$

Here, the left-hand side of the equation denotes the composition of functions, and the right-hand side contains the multiplication of the two matrices and then the application of the product to $z$.

**EXERCISE**: Using simple algebra, check this statement. In other words, let $\gamma = \begin{bmatrix} a & b \\ c & d \end{bmatrix}$ and $\delta = \begin{bmatrix} e & f \\ g & h \end{bmatrix}$, and $\gamma\delta = \begin{bmatrix} p & q \\ r & s \end{bmatrix}$. Show that $\dfrac{a\frac{ez+f}{gz+h}+b}{c\frac{ez+f}{gz+h}+d} = \dfrac{pz+q}{rz+s}$.

For example, it follows from this exercise that the matrix $I$ applied to $z$ yields the neutral motion that sends every point to itself.

(This is very easy to check: If $a = d = 1$ and $b = c = 0$, then $z \to$ $(az + b)/(cz + d) = z/1 = z$.) It also follows that the inverse of a matrix gives the inverse motion.

The group of matrices in $M_2(\mathbf{R})$ with positive determinant[6] is called $\mathrm{GL}_2^+(\mathbf{R})$. So is this group the same as $G^0$? Not quite. We have some built-in redundancy when we represent motions by matrices. Suppose we have

$$\gamma = \begin{bmatrix} a & b \\ c & d \end{bmatrix}.$$

Pick a nonzero real number $\lambda$. Then we can define the new matrix

$$\lambda\gamma = \begin{bmatrix} \lambda a & \lambda b \\ \lambda c & \lambda d \end{bmatrix}.$$

When we form the fractional linear transformation, $\lambda$ cancels out of the numerator and the denominator and we get $\gamma(z) = (\lambda\gamma)(z)$ for all $z$. So $\gamma$ and $\lambda\gamma$, although they will be different matrices if $\lambda \neq 1$, yield the same motion. For example, the matrix $\begin{bmatrix} \lambda & 0 \\ 0 & \lambda \end{bmatrix}$ will yield the neutral motion for any nonzero real number $\lambda$.

---

[6] Because the determinant is multiplicative, if $A$ and $B$ both have positive determinants, so does $AB$. Therefore, $\mathrm{GL}_2^+(\mathbf{R})$ is closed under matrix multiplication, which must be the case for $\mathrm{GL}_2^+(\mathbf{R})$ to be a group with this group law.

*Chapter 12*

# MODULAR FORMS

## 1. Terminology

In mathematical jargon, a function is a rule that associates to every element of some set (the "source") an element of another set (the "target"). Traditionally, certain kinds of functions are called "forms." This happens when the function possesses particular properties. The word "form" also has other meanings—for example, the phrase "space form" is used to denote manifolds having a certain type of shape.

In number theory, the term "form" is sometimes used when a function behaves in a certain way under a change of variables. For example, a polynomial function $f(v)$ is called a "form" of weight $n$ if $f(av) = a^n f(v)$ for every number $a$ (where $v$ can be a vector of one or several variables). We have "quadratic forms" of weight 2, for instance $x^2 + 3y^2 + 7z^2$, or "cubic forms" of weight 3, for instance $x^3 + x^2y + y^3$, and so on. In what follows, we will be concerned with a much more complicated kind of transformational behavior, that of "modular forms."

Modular forms, which we will define soon, are a particular class of "automorphic forms." The word "automorphic" is more descriptive than "modular." Because "auto" means "self" and "morphe" means "shape" in Greek, the adjective "automorphic" is used in contexts where something keeps the same shape under certain changes of variables. This does not mean it stays identically equal to itself under the change of variables, but it does stay close to itself in some sense. The quotient of the transformed function divided by the original function is carefully prescribed, and it is called the "factor of automorphy." When dealing with automorphic forms for the modular group, it is traditional to use the term "modular" instead of "automorphic."

## 2. $SL_2(\mathbf{Z})$

In the previous chapter, we defined a bunch of groups whose members are matrices. This is another one of them. Our number system will be $\mathbf{Z}$, the set of integers. Because we are starting with integers, it is not surprising that the result will be something important to number theory, but the way this happens is very surprising indeed.

Start with $GL_2(\mathbf{Z})$, which we defined in the previous chapter as the set of all 2-by-2 matrices whose entries are integers and whose determinants are $\pm 1$. Here are some elements in $GL_2(\mathbf{Z})$:

$$\begin{bmatrix} 1 & 2 \\ 3 & 7 \end{bmatrix}, \quad \begin{bmatrix} 1 & 1 \\ 0 & 1 \end{bmatrix}, \quad \begin{bmatrix} 1 & 0 \\ 0 & -1 \end{bmatrix}, \quad \begin{bmatrix} 2 & 3 \\ 3 & 4 \end{bmatrix}.$$

The group law is the multiplication of matrices that we defined in the previous chapter. Note that $I = \begin{bmatrix} 1 & 0 \\ 0 & 1 \end{bmatrix}$ is the neutral element in this group, as will be the case for all of the other matrix groups we discuss.

The letter S stands for "special," which in this context means "having determinant equal to 1." So $SL_2(\mathbf{Z})$ is the set of all 2-by-2 matrices with integer entries and determinant 1. It is a subset of $GL_2(\mathbf{Z})$, and because $SL_2(\mathbf{Z})$ has the same group law and the same neutral element as $GL_2(\mathbf{Z})$, we say $SL_2(\mathbf{Z})$ is a *subgroup* of $GL_2(\mathbf{Z})$. Of the four matrices displayed earlier, the first two are in $SL_2(\mathbf{Z})$ and the other two are not.

If necessary, you should review what we said in the previous chapter about fractional linear transformations of the upper half-plane $H$. If

$$\gamma = \begin{bmatrix} a & b \\ c & d \end{bmatrix}$$

is a matrix with real entries and positive determinant, then $\gamma$ gives a function from $H$ to $H$ by the rule

$$z \to \gamma(z) = \frac{az + b}{cz + d}.$$

You can make sure this formula makes sense by working out the imaginary part of the right-hand side to verify that it is positive whenever the imaginary part of $z$ is positive. If we multiply $a$, $b$, $c$, and $d$ by any nonzero real number $\lambda$, we get another matrix that defines the same function from $H$ to $H$.

Let's apply this reasoning to elements of $SL_2(\mathbf{Z})$. We don't want to work with all of $GL_2(\mathbf{Z})$, because the elements of that group with negative determinant map $H$ to the lower half-plane, not to $H$ itself. Any matrix $\gamma$ in $SL_2(\mathbf{Z})$ defines a fractional linear transformation by the rule given earlier, which we write $z \to \gamma(z)$.

There is still a bit of redundancy here in our representation of fractional linear transformations by matrices in $SL_2(\mathbf{Z})$. If we multiply all the entries by the real number $\lambda$, the determinant will be multiplied by $\lambda^2$. Furthermore, the entries are likely to stop being integers. However, if $\lambda = -1$ (and this is the only nontrivial case), we get a new integral matrix of determinant 1 that defines the *same* fractional linear transformation. For instance, $\begin{bmatrix} 1 & 2 \\ 3 & 7 \end{bmatrix}$ and $\begin{bmatrix} -1 & -2 \\ -3 & -7 \end{bmatrix}$ define the same fractional linear transformation, namely

$$z \to \frac{z + 2}{3z + 7} = \frac{-z - 2}{-3z - 7}.$$

Thus, the matrix $-I = \begin{bmatrix} -1 & 0 \\ 0 & -1 \end{bmatrix}$ also defines the do-nothing or neutral motion $z \to z$. We just will live with this redundancy.

This group $SL_2(\mathbf{Z})$ has been studied for a long time and is thoroughly understood. In particular, it has some favorite elements, besides the neutral element $I$. Here they are, with their traditional names:

$$T = \begin{bmatrix} 1 & 1 \\ 0 & 1 \end{bmatrix}, \qquad S = \begin{bmatrix} 0 & 1 \\ -1 & 0 \end{bmatrix}.$$

We've actually seen the motions springing from $S$ and $T$ before. Namely, $T(z) = z + 1$, so $T$ simply shifts the whole upper half-plane over by one unit, and $S(z) = -1/z$ is the inversion we described in the previous chapter.

Here's a nice theorem:

**THEOREM 12.1**: *The group* $SL_2(\mathbf{Z})$ *is generated by* $S$ *and* $T$.

This theorem means that any element of the group can be obtained by multiplying $S$'s, $T$'s, $S^{-1}$'s, and $T^{-1}$'s together. This fact is closely related to the theory of continued fractions, but we won't go in that direction here. A careful analysis may be found in Series (1985).

Notice that $S^2 = -I$ so $S^4 = I$, but no power of $T$ equals $I$, except for the zeroth power. In fact, you can multiply matrices together to check that

$$T^k = \begin{bmatrix} 1 & k \\ 0 & 1 \end{bmatrix}.$$

This formula is true for positive $k$, for $k = 0$ (because we define the zeroth power of a group element to be equal to the neutral element), and for negative $k$ (because we define $g^{-m}$ to be the product of $m\, g^{-1}$'s, if $m > 0$ and $g$ is a group element).

We say that $SL_2(\mathbf{Z})$ *acts* on the upper half-plane $H$. This statement means that each element of this group defines a function $H \to H$, and certain rules apply. Namely,

(1) For any $z$ in $H$ and any matrices $g$ and $h$ in $SL_2(\mathbf{Z})$, we have $g(h(z)) = (gh)(z)$, and

(2) If $I$ is the neutral element in $SL_2(\mathbf{Z})$, then $I(z) = z$.

## 3. Fundamental Domains

Start with your favorite point $z_0$ in $H$. (For example, $z_0$ might be $i$.) If you look at everywhere $z_0$ goes when you act on $z_0$ by all the matrices in $SL_2(\mathbf{Z})$, then you get a subset of points of $H$ that is called the *orbit* of $z_0$. Because this is an action by motions that preserve the distance in $H$, these points cannot bunch up on each other.[1] We say that the action is *discrete*.

The discreteness of the action is what makes the whole game of modular forms playable. As we will see—sooner or later—a modular form is a kind of function whose values on all the points of a single orbit are closely related. If those points bunched up, then because of the laws of complex analysis, there couldn't be any interesting modular forms.

NOTE: There are other groups we could look at, alongside $SL_2(\mathbf{Z})$. We could look at *congruence subgroups* of $SL_2(\mathbf{Z})$, which we will define in chapter 14.

The last thing in this section is to draw for you a couple of fundamental domains for the action of the modular group on $H$. (You can look back at the previous chapter for a different example of the concept of fundamental domain.) A fundamental domain for this action is a subset $\Omega$ of $H$ with two properties:

(1) Every orbit intersects $\Omega$.
(2) Two different points of $\Omega$ cannot be in the same orbit.

In symbols, we can say equivalently that

(1) For any $z_1$ in $H$, there exists some $z_0$ in $\Omega$ and some $\gamma$ in $SL_2(\mathbf{Z})$ with the property that $z_1 = \gamma(z_0)$.
(2) If $z_2$ and $z_3$ are in $\Omega$, and if $\gamma$ is in $SL_2(\mathbf{Z})$, and if $z_3 = \gamma(z_2)$, then $z_3 = z_2$.

There are many different fundamental domains. If $\Omega$ is a fundamental domain, then so is $\gamma(\Omega)$ for any $\gamma$ in $SL_2(\mathbf{Z})$. But you could also do weirder things, like chop up $\Omega$ into pieces and move them around in various ways. Naturally, we like to deal with nice-looking fundamental domains.

---

[1] If you plot an orbit, the points will *appear* to bunch up near the real axis, but that's because of an optical illusion: The Euclidean distance on the paper is not the same as the non-Euclidean distance defined by that integral in section 4 of the previous chapter. Far from it.

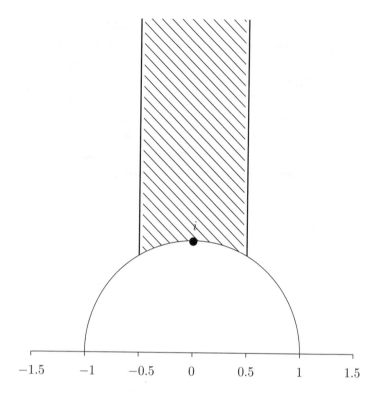

Figure 12.1. The standard fundamental domain $\Omega$

The "standard" fundamental domain $\Omega$ consists of points $x + iy$ with $-0.5 \leq x < 0.5$ and $x^2 + y^2 > 1$, along with the piece of the circle $x^2 + y^2 = 1$ with $-0.5 \leq x \leq 0$. It is the shaded region in figure 12.1, if we are careful about which parts of the boundary are included.

We can take the region in figure 12.1 and apply the function $z \mapsto -1/z$ mentioned above. That gives a different fundamental region, shaded in figure 12.2, where now the boundaries of the region (not specified in the figure) are pieces of three different semicircles. If you are careful, you can work out which pieces of these semicircles correspond to the boundaries in figure 12.1.

Notice that the second fundamental domain comes down to 0 on the real line in a pointy way. For that reason, we call 0 a "cusp"— it is the cusp of that fundamental domain. But 0 is not *in* the

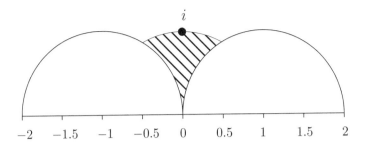

Figure 12.2. A different fundamental domain

fundamental domain, because the real line is not part of the upper half-plane.

## 4. Modular Forms at Last

Recall the concept of an "analytic function" $f(z)$ from section 4 of chapter 7. This is a complex-valued complex-differentiable function of a complex variable $z$ with domain an open subset $U$ of $\mathbf{C}$. (We defined "open set" there.) A theorem of complex analysis states that an analytic function $f$ has a convergent Taylor series around each point where it is defined (although the radius of convergence may vary from point to point), and that each of these Taylor series converges to $f$ within its disc of convergence.

In other words, an analytic function $f : U \rightarrow \mathbf{C}$, where $U$ is a set such as the upper half-plane $H$, punctured unit disc $\Delta^*$, or unit disc $\Delta^0$, can be represented as a power series around each point of $U$. Namely, for any point $z_0$ in $U$, we can write

$$f(z) = a_0 + a_1(z - z_0) + a_2(z - z_0)^2 + a_3(z - z_0)^3 + \cdots,$$

where the infinite series converges for all $z$ sufficiently close to $z_0$ to a limit that equals the value of the function, $f(z)$.

It is not too hard to see that constant functions are analytic, polynomial functions are analytic, and the sum, difference, and product of analytic functions are analytic. The quotient of two analytic functions is also analytic, as long as the denominator doesn't vanish at any point of $U$. The composition of analytic functions is also analytic. Basically, the usual rules of calculus apply to analytic functions.

Okay, so a modular form is a certain kind of analytic function $f : H \to \mathbf{C}$. We are going to define a "modular form of level 1 and weight $k$," where $k$ is a nonnegative even integer. This restriction helps us to keep things simple. The words "level 1" mean that $f$ behaves nicely under *all* the fractional linear transformations coming from $\mathrm{SL}_2(\mathbf{Z})$. If instead we demand nice behavior only under some smaller subgroup of $\mathrm{SL}_2(\mathbf{Z})$ (namely a *congruence subgroup*), we still get an interesting theory, but with larger "level." If we make the level higher, we can also make the weight $k$ odd, or even fractional. Fractional weight is particularly hairy, but also important in the theory, and we will mention fractional weight forms at one or two points later.

But unless we specify further, we will just use the term "modular form of weight $k$" as defined here. It is quite amazing that such functions could have wrapped up in them so much deep number theory. Although they were invented a couple of centuries ago, the properties and uses of modular forms in number theory only gradually came to light over the years.

We keep beating about the bush. It is time to give the definition:

**DEFINITION**: Let $k$ be a nonnegative even integer. A *modular form of weight $k$* is an analytic function $f : H \to \mathbf{C}$ on the upper half-plane with two properties:

(1) Transformation property.
(2) Growth property.

It is very important to note that, from the outset, we require $f$ to be analytic. This restriction is not absolutely necessary. We could allow $f$ to have poles—points where $f$ goes to infinity in a controlled way—and there is a theory of a kind of modular form that is not even complex-differentiable at all, but we won't pursue that.

Let's discuss the two numbered properties one at a time.

## 5. Transformation Property

The transformation property is a kind of periodicity of the function that is the modular form. To explain, first suppose that $X$ is a real variable. A periodic function $g(X)$ of period 1 satisfies a "functional equation": $g(X+1) = g(X)$. We have seen some examples of periodicity in previous chapters. We could vary this by asking for a different functional equation. For instance, we could ask that $g(X+1) = \phi(X)g(X)$, where $\phi(X)$ is some known, given function, called an *automorphy factor*.[2] Now the values of $g(X)$ do not simply repeat as we slide $X$ one unit to the right, but still they are predictable from the earlier values. If we know the function $\phi(X)$ for all $X$, and if we know $g(A)$ for every $A$ from 0 to 1, then we can figure out $g(X)$ for all $X$. This is no longer periodicity. Instead, we could call it "automorphy," so long as $\phi$ is a relatively simple function.

In the nineteenth century, some mathematicians found that the automorphic transformation property we are about to set down helped them solve problems and led to beautiful mathematics. Let $\gamma$ be a matrix in $\mathrm{SL}_2(\mathbf{Z})$:

$$\gamma = \begin{bmatrix} a & b \\ c & d \end{bmatrix}.$$

Remember that $a, b, c,$ and $d$ are integers and that the determinant $\det(\gamma) = ad - bc = 1$. Also, remember that $\gamma$ defines a function from $H$ to $H$ by the rule

$$\gamma(z) = \frac{az+b}{cz+d}.$$

Just for fun, let's work out the derivative of $\gamma(z)$ using the usual rule for the derivative of a quotient.

$$\frac{d}{dz}(\gamma(z)) = \frac{(az+b)'(cz+d) - (az+b)(cz+d)'}{(cz+d)^2}$$

$$= \frac{a(cz+d) - (az+b)c}{(cz+d)^2} = \frac{ad-bc}{(cz+d)^2} = (cz+d)^{-2}.$$

---

[2] For example, suppose $g(X) = e^X$. Then $g(X+1) = e^{X+1} = ee^X = eg(X)$. In this simple example, the automorphy factor $\phi(X)$ is the constant function $e$.

The derivative simplified quite nicely. Because the derivative is an important object, it is to be expected that the expression $(cz + d)^{-2}$ will be important in our theory. That the exponent is $-2$ is significant: It means that our theory works best for even weights. The weight is the integer $k$ we specified in our definition.

Here is the transformation property: For any $\gamma$ in $\mathrm{SL}_2(\mathbf{Z})$ with entries $a$, $b$, $c$, and $d$ as earlier, and for all $z$ in the upper half-plane $H$, the transformation property for a modular form of weight $k$ is

$$f(\gamma(z)) = (cz + d)^k f(z). \tag{12.2}$$

The connection between (12.2) and the derivative of $\gamma(z)$ is clear on a formal algebraic level for $k = 2$, but its deeper meaning, connected with the geometry of higher dimensional manifolds called "vector bundles over the upper half-plane" (cryptically referred to on page 177), is unfortunately beyond our scope. However, you will see later that this is a fruitful definition, at least because of its utility in uncovering very surprising patterns in the number theory that we have been discussing in this book.

Right away, you should realize that if $f$ satisfies (12.2) for all $\gamma$ in $\mathrm{SL}_2(\mathbf{Z})$, and if you know the values of $f$ on a fundamental domain $\Omega$, then you know the values of $f$ everywhere. This is because $cz + d$ is a completely explicit, computable factor. And if $z$ is any point in $H$, we can find a unique point $z_0$ in $\Omega$ and an element $\gamma$ in $\mathrm{SL}_2(\mathbf{Z})$ with the property that $z = \gamma(z_0)$. Then the transformation law tells us that $f(z) = (cz_0 + d)^k f(z_0)$. We know $f(z_0)$ and we know $cz_0 + d$, so we know $f(z)$.

Another thing to check is that if (12.2) is true for a couple of elements of $\mathrm{SL}_2(\mathbf{Z})$, say $\gamma$ and $\delta$, then it is also true for their product $\gamma\delta$. This is a calculation we leave as an exercise for you. It is related to the chain rule and the fact that the factor of automorphy $cz + d$ is the derivative of the fractional linear transformation. Similarly, if (12.2) is true for a matrix $\gamma$ in $\mathrm{SL}_2(\mathbf{Z})$, then it is also true for $\gamma^{-1}$.

Now we said earlier that $T$ and $S$ generate the whole group $\mathrm{SL}_2(\mathbf{Z})$. It follows from this assertion and the preceding paragraph that if we merely know that the functional equation is true for $T$

and for $S$, then we know it holds for every element of $\mathrm{SL}_2(\mathbf{Z})$. So we could replace (12.2) as stated earlier with the two simpler-looking laws

$$f(z+1) = f(z) \quad \text{and} \quad f(-1/z) = z^k f(z).$$

It is better to stick with (12.2). Sometimes, we work not only with the group $\mathrm{SL}_2(\mathbf{Z})$ but also with various subgroups that are not generated only by the two matrices $T$ and $S$. These subgroups have their own list of generators. There could be millions of generators, and they depend on which subgroup we work with. A modular form for one of these smaller groups would only be required to satisfy the functional equation for $\gamma$'s in the smaller group.

But we can run immediately with the first of these special simpler-looking laws. Because $f(z+1) = f(z)$, we see that $f(z)$ is periodic of period 1. From chapter 8, we know that this means that we can write $f(z)$ as a function of $q = e^{2\pi i z}$. Because $f(z)$ is analytic, we can prove that it can be written in terms of a convergent power series in $q$.

But there is a wrinkle here: Remember that $z \to q$ sends $H$ to the punctured disc $\Delta^*$. This disc is *punctured*. So $f(z)$, which can also be thought of as a function of $q$ running around $\Delta^*$, will be an analytic function $F(q)$ (say) of $q$, but one that has not been defined at the center 0 of the disc. The upshot of this is that while $F(q)$ can be described by a power series in $q$, this series may contain negative powers of $q$ as well as positive powers. In fact, at this point, there could even be *infinitely* many negative powers of $q$ appearing. If there are negative powers of $q$, then the series is not a Taylor series around 0. But since $F(q)$ is not defined at 0, we can't expect it necessarily to have a Taylor series there.

To sum up, from the fact that $f(z)$ is analytic on $H$ and satisfies the functional equation for all $\gamma$ in $\mathrm{SL}_2(\mathbf{Z})$, we know that there are constants $a_i$, for $i$ running over all integers, such that

$$f(z) = \cdots + a_{-2}q^{-2} + a_{-1}q^{-1} + a_0 + a_1 q + a_2 q^2 + \cdots,$$

where $q = e^{2\pi i z}$. We call the right-hand side the "$q$-expansion" of $f(z)$.

# 6. The Growth Condition

If we left things as they are, there would be too many modular forms. We would have chaos, and we wouldn't be able to say nice things about all modular forms. The problem is all those negative exponents in the $q$-expansion. If they are really there, then the modular form will blow up to infinity or behave even more wildly as $q \to 0$ (which is when the imaginary part of $z$ tends to $\infty$).

If there are *no* negative exponent terms in the $q$-expansion of $f(z)$, then we say $f(z)$ is a *modular form*. We can state our growth condition then as follows:

GROWTH CONDITION: $|f(z)|$ remains bounded as the imaginary part of $z$ tends to $\infty$.

In this case, we can actually ascribe a value to the limit of $f(z)$ as the imaginary part of $z$ tends to $\infty$, because this limit will be the same as the limit of the $q$-expansion as $q$ tends to 0. When there are no negative exponent terms, this limit exists and equals $a_0$.

For future reference, we define "cusp form" as follows:

**DEFINITION**: The modular form $f(z)$ is a *cusp form* if the value of $a_0$ in its $q$-expansion equals 0.

This definition is clearly equivalent to saying that the value of the $q$-expansion tends to 0 as $q \to 0$. If you look at the second fundamental domain in our picture, you see that $q \to 0$ is geometrically related to $z$ tending to a cusp while remaining inside of $\Omega$. That's why the term "cusp form" is used.

# 7. Summary

A *modular form* of weight $k$ is an analytic function on $H$ satisfying (12.2) for all $\gamma$ in the modular group and that possesses a $q$-expansion of the form

$$f(z) = a_0 + a_1 q + a_2 q^2 + \cdots .$$

We have a symbol for the set of all modular forms of weight $k$, namely $M_k$.

A *cusp form* of weight $k$ is an analytic function on $H$ satisfying (12.2) for all $\gamma$ in the modular group and that possesses a $q$-expansion of the form

$$f(z) = a_1 q + a_2 q^2 + \cdots.$$

The symbol for the set of all cusp forms of weight $k$ is $S_k$. (The S is for *Spitze*, the German word for cusp.)

By the way, if the $q$-expansion of the modular form $f(z)$ has only a finite number of *positive* terms (i.e., it is a polynomial), then it can be proven that $f(z)$ must be identically 0.

An example of a modular form of weight 4 is the function $G_4 : H \to \mathbf{C}$, defined by

$$G_4(z) = \sum_{\substack{m=-\infty \\ }}^{\infty} \sum_{\substack{n=-\infty \\ (m,n) \neq (0,0)}}^{\infty} \frac{1}{(mz+n)^4}.$$

This example is explained in the next chapter, and many more examples of modular forms will also be given in later chapters.

**EXERCISE**: Prove that if $f(z)$ is a modular form (for the full group $SL_2(\mathbf{Z})$, as we have been assuming in this chapter) of weight $k > 0$, where $k$ is an *odd* integer, then $f(z) = 0$ for all $z$ in the upper half-plane (i.e., $f(z)$ is the 0 function). HINT: Think about (12.2) when $\gamma = -I$. We will work out the details in section 1 of chapter 14.

# Chapter 13

# HOW MANY MODULAR FORMS ARE THERE?

## 1. How to Count Infinite Sets

We continue this chapter directly from the previous chapter. Fix a nonnegative even integer $k$. For example, you can take $k = 12$. We defined what it means for $f(z)$ to be a modular form of weight $k$, or a cusp form of weight $k$, and we fixed the notations $M_k$ for the set of the former and $S_k$ for the set of the latter.[1]

How big are $M_k$ and $S_k$? We can see right away that each of these sets has either one element exactly or has infinitely many elements. Why? Well, the constant function 0 is always a modular form of weight $k$ (for any weight). This may be a trivial observation, but it is unavoidable. The function $z \to 0$ satisfies all the properties needed to be a modular form. In fact, it is a cusp form. So $M_k$ and $S_k$ each possess at least one element, namely 0.

Now suppose there is a nonzero element, $f(z) \neq 0$ in $M_k$, say. If $c$ is any complex number whatever, then $cf(z)$ is also a modular form of weight $k$. (Check the definitions—you can multiply the formulas through by $c$.) So one nonzero element gives you infinitely many.

Another thing you can do with $M_k$ and $S_k$ is to add modular forms together. If, for example, $f(z)$ and $g(z)$ are both modular forms of weight $k$, then so is $f(z) + g(z)$. (Check the definitions again. This verification requires that $f$ and $g$ share the *same* weight, so you can pull out the automorphy factor in the transformation law of the sum, using the distributive law of multiplication.)

---

[1] REMINDER: Remember that $M_k$ and $S_k$ are sets of modular forms of level 1: modular forms that transform in the right way under *all* fractional linear transformations coming from the group $\mathrm{SL}_2(\mathbf{Z})$.

We have a name for this sort of gadget. If $V$ is a set of functions from $\Sigma \to \mathbf{C}$ (here $\Sigma$ can be *any* set), then we say that $V$ is a *vector space* if

(V1) $V$ is nonempty.

(V2) For any function $v$ in $V$ and any complex number $c$, the function $cv$ is also in $V$.

(V3) For any functions $v$ and $w$ in $V$, the function $v + w$ is also in $V$.

If $V$ is a vector space, then by putting rules (V2) and (V3) together and iterating them, you can see that if $v_1, v_2, \ldots, v_n$ are all in $V$ and $c_1, c_2, \ldots, c_n$ are all complex numbers, then

$$v_1 c_1 + v_2 c_2 + \cdots + v_n c_n \tag{13.1}$$

is also in $V$. We call (13.1) a *linear combination* of the functions $v_1$, $v_2, \ldots, v_n$.

We may sometimes call the elements of a vector space "vectors." This terminology is useful because we could derive a whole theory of vector spaces starting not with functions but with an abstract set of objects that satisfy rules (V1), (V2), and (V3). This theory is called "linear algebra." In this context, we often speak of complex numbers as *scalars* and we say that $cv$ is a *scalar multiple* of $v$ if $c$ is a scalar and $v$ is a vector.

From our discussion so far, we can say $M_k$ is a vector space. What about $S_k$? It is a subset of $M_k$, but that's not enough for it to be a vector space on its own. It still must satisfy rules (V1), (V2), and (V3). Because 0 is a cusp form, rule (V1) is satisfied. Now, what makes a cusp form a cusp form? The constant term of its $q$-expansion has to equal 0. Because $c0 = 0$ for any complex number $c$, and $0 + 0 = 0$, we see that rules (V2) and (V3) are also satisfied for $S_k$. Because $S_k$ is a vector space on its own *and* a subset of $M_k$, we say that $S_k$ is a *subspace of $M_k$*.

Now let's consider any vector space $V$. Our reasoning for $M_k$ and $S_k$ carries over to $V$, and we learn that either $V = \{0\}$ and has exactly one element, namely the 0 function, or else $V$ is an infinite set of functions. In any case, whenever $V$ possesses some elements, it also possesses all the linear combinations we can make out of them.

There now comes an important dichotomy. *Either*

(1) There is a finite set $S$ of vectors in $V$ whose linear combinations exhaust all of $V$

    *or*

(2) There is no such set.

**EXAMPLE**: If $V = \{0\}$, then we can take $S$ to be $V$ itself. However, it is important to note that we can even take $S$ to be the empty set. That's because we make the convention that 0 is a linear combination of *no* vectors. (Compare chapter 10, section 1.)

**EXAMPLE**: If $v$ is a nonzero function, then we can form the set $V = \{cv \mid c \in \mathbf{C}\}$. In words, $V$ is the set of all scalar multiples of $v$. You can check rules (V1), (V2), and (V3), again by using the distributive law, and see that $V$ is a vector space. In this case, we can take the "generating set" $S$ to be the singleton $\{v\}$. However, we could be sloppy and take a larger $S$, such as $\{v, 3v, 0, (1 + i)v\}$.

**EXAMPLE**: Let $V$ be the set of all polynomial functions of $z$. A generating set $S$ for this $V$ could not possibly be finite, because the degree of a linear combination of a finite set $S$ of polynomials could never exceed the highest degree of any of the members of $S$. Even so, the set of all polynomials satisfies the properties of a vector space.

Vector spaces of the first kind are called "finite-dimensional," and those of the second kind are called—surprise—"infinite-dimensional."

Now suppose we have a finite-dimensional vector space $V$. The vector space will have various generating sets $S$, which will have various sizes. Some of these generating sets are finite, because of the definition of "finite-dimensional." Of all these finite generating sets, some of them will be minimal. (A generating set $S$ of $V$ is *minimal* if none of its proper subsets generate $V$.) We call any one of these minimal generating sets a *basis* for $V$.

For example, the 0-vector space has the empty set for its basis. This example is the only time there is no choice involved in finding a basis. Our second example, $V = \{cv \mid c \in \mathbf{C}\}$, has for a basis the singleton set $\{v\}$. (Clearly this is the smallest possible generating set of vectors for this $V$.) But $\{3v\}$ is another basis for the same $V$. Notice that $\{3v\}$ is also a single-element set.

Here is the nice theorem that is proved in a course on linear algebra:

> **THEOREM 13.2**: *Let $V$ be a finite-dimensional vector space. Then all of its bases have the same number of elements as each other.*

If the number of elements in a basis is $d$, then $d$ is called the *dimension* of $V$. We write $\dim(V) = d$. Another theorem says that if $W$ is a subspace of $V$, then $\dim(W) \le \dim(V)$.

In our two examples so far, the 0-vector space has dimension 0 (because the empty set, which is its basis, has 0 elements). The vector space $V = \{cv \mid c \in \mathbf{C}\}$, where $v \ne 0$, has dimension 1.

> **EXERCISE**: Fix some positive integer $n$. What is the dimension of the vector space of all polynomial functions of $z$ of degree less than or equal to $n$?

> **SOLUTION**: The answer is $n + 1$. A basis of this vector space is $\{1, z, z^2, \ldots, z^n\}$.

The notion of dimension is how we can talk sensibly about how many modular forms there are of a given weight. We can say whether or not $M_k$ or $S_k$ is a finite-dimensional vector space and, if so, what its dimension is. In this way, it makes sense to say one vector space is "bigger" than another, even though both may contain infinitely many elements.

## 2. How Big Are $M_k$ and $S_k$?

Before rushing onto the answer, let's see if there is anything we can say *a priori*. We don't know whether there are any nonzero modular forms of any weight at this point. But it is always true that $S_k$ is a subspace of $M_k$, so $\dim(S_k) \leq \dim(M_k)$. How large could $\dim(M_k) - \dim(S_k)$ be? At this point, for all we know, every modular form might be a cusp form, so the difference could be 0. Another possibility is that the difference is 1. Let's now see why the difference could never be greater than 1, assuming that $M_k$ is a finite-dimensional vector space.

Suppose $\{f_1, \ldots, f_n\}$ is a basis of $M_k$, so $\dim(M_k) = n$. Each one of these functions has a $q$-expansion, which has a constant term. Let's set $b_i$ to be the constant term of the $q$-expansion of $f_i$. So $b_i$ is some complex number. If all of the $b_i$ are zero, then all of the $f_i$ are cusp forms, and $M_k = S_k$. If some of the numbers $b_i$ are not zero, we may as well assume that $b_1 \neq 0$. Now let's create some new functions in $M_k$,

$$g_i = f_i - (b_i/b_1)f_1,$$

for $i = 2, 3, \ldots, n$. We leave it to you to see that $\{f_1, g_2, g_3, \ldots, g_n\}$ is also a generating set of $M_k$.

What is the constant term of the $q$-expansion of $g_i$? It is the constant term of the $q$-expansion of $f_i$ minus $(b_i/b_1)$ times the constant term of the $q$-expansion of $f_1$. This number is $b_i - (b_i/b_1)b_1 = 0$. Therefore all the $g_i$ are cusp forms. Because $\{f_1, \ldots, f_n\}$ is a minimal generating set of $M_k$, it's not hard to see that $\{g_2, \ldots, g_n\}$ forms a minimal generating set of $S_k$. (The precise proof depends on using the notion of *linear independence*.) So in this case $\dim(S_k) = n - 1$.

In summary, we either have $M_k = S_k$, so the two vector spaces have the same dimension, or else $M_k$ is strictly larger than $S_k$ and its dimension is exactly bigger by 1.

Now it is time to see why there are nonzero modular forms. How could we construct one? There is only one semiobvious thing to do, which leads to the construction of what are called "Eisenstein series." We will see that for every even integer $k \geq 4$ we can write down explicitly a modular form of weight $k$ that is not zero. (The exercise at the end of chapter 12 tells us[2] that there is no point in considering odd values of $k$.) In fact, the functions we will now define are not cusp forms.

The idea is to start with the automorphy factor itself: $z \to (cz + d)^{-k}$. Because there is only one type of automorphy factor, this gives us only one idea. There are infinitely many ordered pairs $(c, d)$, so which values should we use? We'll have to use them all. Now $(c, d)$ is the bottom row of a matrix in $\mathrm{SL}_2(\mathbf{Z})$. So $c$ and $d$ are relatively prime integers, because $ad - bc = 1$. (Conversely, if $c$ and $d$ are relatively prime integers, then there exists another pair of integers $a$ and $b$ such that $ad - bc = 1$. This is theorem 1.1.) However, it turns out that we gain in simplicity if we allow $c$ and $d$ to be nearly any pair of integers in the expression $(cz + d)^{-k}$. Of course, we cannot use $(0, 0)$, because then we would be dividing by 0, but we allow every other possible pair of integers.

Let $\Lambda'$ stand for the set of all ordered pairs of integers $(c, d)$ excluding $(0, 0)$. The reason for this notation is that $\Lambda$ is often used to stand for the set of all ordered pairs of integers, and we want a subset of $\Lambda$.

OKay, let's try to create a modular form, which we will call $G_k(z)$. First, let's just try a particular $(c, d)$. Why not $(1, 0)$? We are thus looking at the function $z \to 1/z^k$. Perhaps this is a modular form of weight $k$? This function is analytic on the upper half-plane. What about its $q$-expansion? Whoops, it doesn't have a $q$-expansion, because it is not periodic of period 1 (or any other period).

To remedy this deficiency, we can create a periodic function out of $z \to 1/z^k$ in the following way. We define a new function:

$$ h_k(z) = \sum_{n=-\infty}^{\infty} \frac{1}{(z+n)^k}. $$

This formula may be brutal, but the right-hand side is visibly unchanged if you replace $z$ by $z + 1$. But we've paid a price. The right-hand side is an infinite sum of analytic functions. If $k = 0$, for example, the right-hand side adds up to infinity, which is no good.

---

[2] REMEMBER: Right now we are only talking about modular forms of level 1.

However, if the right-hand side converges for every $z$ in $H$, then it will define a function of $z$, and if that convergence is uniform on compact sets, that function will again be analytic. For the moment, let's suppose the right-hand side is a convergent series that adds up to an analytic function of $z$ in $H$. Could $h_k(z)$ be a modular form?

First, let's look at the second requirement of modularity—the growth requirement. You can skip the following argument, which is not rigorous anyway. But if you are interested, it will give you the flavor of how some of these growth estimates work. The more complicated growth estimates needed in other proofs are similar, but we will skip them entirely.

We want to look at the $q$-expansion of the periodic function $h_k(z)$. We don't really want to figure out what it is exactly. We just want to see how $h_k(z)$ grows when $q \to 0$. This is the same as seeing how $h_k(z)$ grows when $z = x + iy$ tends to $\infty$ while staying in the fundamental domain $\Omega$ from figure 12.1. In other words, $-1/2 \leq x < 1/2$ and $y \to \infty$.

By the triangle inequality, we know that

$$|h_k(z)| \leq \sum_{n=\infty}^{\infty} \left| \frac{1}{(z+n)^k} \right|.$$

Work out

$$\left| \frac{1}{(z+n)^k} \right| = \frac{1}{|z+n|^k} = \frac{1}{|(x+n+iy)|^k} = \frac{1}{((x+n)^2+y^2)^{k/2}}.$$

Now, for a fixed $x$, when we sum over $n$, most of the terms have $n$ much larger than $x$ and the particular value of $x$ is irrelevant. So we can say that the growth of $|h_k(z)|$ for $z$ in $\Omega$ as $y \to \infty$ will be the same as the growth of

$$\sum_{n=-\infty}^{\infty} \frac{1}{(n^2+y^2)^{k/2}}.$$

It's a fair guess that this sum is about the same size as the integral

$$\int_{t=-\infty}^{\infty} \frac{dt}{(t^2+y^2)^{k/2}}.$$

For a fixed $y$, you can do the definite integral explicitly. Assuming $k \geq 2$ and $y \neq 0$, it works out to be $C_k/y^{p_k}$ for some positive constant $C_k$ and some positive integer $p_k$. Therefore, the limit as $y \to \infty$ is 0, which is certainly finite. So the growth condition is satisfied.

Last, we have to look at how $h_k(z)$ transforms under $\gamma = \begin{bmatrix} a & b \\ c & d \end{bmatrix}$ in $SL_2(\mathbf{Z})$. We compute:

$$h_k\left(\frac{az+b}{cz+d}\right) = \sum_{n=-\infty}^{\infty} \frac{1}{\left(\frac{az+b}{cz+d}+n\right)^k}.$$

Let's work out one of the terms on the right:

$$\frac{1}{\left(\frac{az+b}{cz+d}+n\right)^k} = \frac{1}{\left(\frac{(az+b)+n(cz+d)}{cz+d}\right)^k} = \frac{(cz+d)^k}{((a+nc)z+(b+nd))^k}.$$

Now we want the transformation to follow our law in the definition of modular form. So we should pull out the automorphy factor to better understand what we have. Pulling it out and adding up all the terms gives

$$h_k\left(\frac{az+b}{cz+d}\right) = (cz+d)^k \sum_{n=-\infty}^{\infty} \frac{1}{((a+nc)z+(b+nd))^k}.$$

What we are hoping for (if we were to have a modular form) is

$$h_k\left(\frac{az+b}{cz+d}\right) \stackrel{?}{=} (cz+d)^k h_k(z).$$

At least we get the automorphy factor coming out right. Of course, that's why we started with it in the first place. But the new infinite sum is summing over wrong things, although they have the right shape.

Maybe we are not summing over enough terms? Starting with $h_k(z)$ and transforming by $\gamma$, we get the $k$th powers of new linear terms in the denominators of the sum, terms where the coefficient of $z$ is not 1 but some other integer. We've already got our feet wet summing up an infinite number of terms, so why not go all the way? We define

$$G_k(z) = \sum_{(m,n)\in\Lambda'} \frac{1}{(mz+n)^k}.$$

This formula turns out to work, as we shall see soon. It is called the *Eisenstein series* of weight $k$ (for the full modular group).

Now it is time to worry about the convergence of the infinite series. We are summing up a lot more terms than before. On the other hand, the larger the value of $k$, the smaller most of the terms will be. It turns out that our formula does not work well for $k = 0$ or $k = 2$. From now on, when discussing Eisenstein series, we will assume the weight $k$ is 4 or greater. In that case, the series converges absolutely uniformly on compact subsets of $H$, and we don't have to worry about the order of summation of the terms of the double sum over $m$ and $n$.

Each term is an analytic function on $H$, and the series converges absolutely uniformly on compact sets. Complex analysis tells us that $G_k(z)$ is an analytic function on $H$. Estimates similar to what we did for $h_k(z)$ show that $G_k(z)$ satisfies the growth requirement. All we have left to do is to check the transformation law under $\gamma = \left[\begin{smallmatrix} a & b \\ c & d \end{smallmatrix}\right]$ in $\mathrm{SL}_2(\mathbf{Z})$. Again, we compute:

$$G_k\left(\frac{az+b}{cz+d}\right) = \sum_{(m,n)\in\Lambda'} \frac{1}{\left(m\left(\frac{az+b}{cz+d}\right)+n\right)^k}.$$

Let's again work out one of the terms on the right:

$$\frac{1}{\left(m\left(\frac{az+b}{cz+d}\right)+n\right)^k} = \frac{1}{\left(\frac{m(az+b)+n(cz+d)}{cz+d}\right)^k}$$

$$= \frac{(cz+d)^k}{((ma+nc)z+(mb+nd))^k}.$$

Summing it up,[3] we get

$$G_k\left(\frac{az+b}{cz+d}\right) = (cz+d)^k \sum_{(m,n)\in\Lambda'} \frac{1}{((ma+nc)z+(mb+nd))^k}.$$

Now comes a key step. In looking at the sum on the right-hand side of this equation, remember that $a$, $b$, $c$, and $d$ are constants and $m$ and $n$ are the variables in the summation. We can define a change of variables as follows: $m' = ma + nc$ and $n' = mb + nd$.

---

[3] Pun intended.

We claim that this change of variables determines a one-to-one correspondence from $\Lambda'$ to itself.[4] In other words, we can rewrite the previous equality as

$$G_k\left(\frac{az+b}{cz+d}\right) = (cz+d)^k \sum_{(m',n')\in\Lambda'} \frac{1}{(m'z+n')^k}.$$

But $m'$ and $n'$ are just dummy summation variables, no different from $m$ and $n$. The sum on the right is just $G_k(z)$ again. We conclude that

$$G_k\left(\frac{az+b}{cz+d}\right) = (cz+d)^k G_k(z).$$

This is the desired transformation law for a modular form of weight $k$.

## 3. The $q$-expansion

Perhaps it is not totally surprising that the Eisenstein series will have something to do with sums of powers of integers. This discovery appears when we find their $q$-expansions. We won't prove anything about this here, but we will tell you the answer. Remember that $G_k(z)$ has a $q$-expansion because it is analytic and periodic of period 1. To write down its $q$-expansion, we have to remind you about some functions and constants that we defined in earlier parts of this book, and introduce a new function as well.

(1) The $\zeta$-function:

$$\zeta(s) = \sum_{n=1}^{\infty} \frac{1}{n^s}.$$

If $k$ is a positive even integer (as it has been in the current section), then $\zeta(k)$ is the infinite sum of the reciprocals of the $k$th powers of the positive integers.

---

[4] Because $ad - bc = 1$, you can solve those equations for $m$ and $n$ in terms of $m'$ and $n'$, which is precisely the way to check that this is a one-to-one correspondence. You also need to check that when $m = n = 0$, then $m' = n' = 0$ and vice versa.

(2) The divisor-power-sum functions, for any nonnegative integer $m$:

$$\sigma_m(n) = \sum_{d\mid n} d^m.$$

The sum is over all the positive divisors $d$ of the positive integer $n$, including 1 and $n$ itself.

(3) The Bernoulli numbers $B_k$, which were related to the sums of consecutive $k$th powers.

With these three items, we can tell you the $q$-expansion of the Eisenstein series $G_k$,

$$G_k(z) = 2\zeta(k)\left(1 - \frac{2k}{B_k}\sum_{m=1}^{\infty}\sigma_{k-1}(n)q^n\right), \tag{13.3}$$

where, as usual, $q = e^{2\pi i z}$.

Formula (13.3) is pretty amazing. We can see where the factor of $\zeta(k)$ comes from. Recall the definition of the Eisenstein series:

$$G_k(z) = \sum_{(m,n)\in\Lambda'} \frac{1}{(mz+n)^k}.$$

As long as $m$ and $n$ are not both 0, we can compute their greatest common divisor $d$. Because we are assuming by now that $k \geq 4$, the sum for $G_k$ is absolutely convergent, and we can compute the sum in any order. So we can partition all the pairs in $\Lambda'$ by placing together in a single part all those with the same greatest common divisor (GCD). In the part with GCD equal to $d$, we can write all those $(m,n)$'s as $(da, db)$, where $(a, b)$ are relatively prime (i.e., have greatest common divisor 1). In this way, we can write

$$G_k(z) = \sum_{d=1}^{\infty}\frac{1}{d^k}\sum_{(a,b)\in\Lambda''}\frac{1}{(az+b)^k},$$

where $\Lambda''$ is the subset of $\Lambda$ consisting of all pairs of relatively prime integers. Because $\Lambda''$ does not depend on $d$, we were able to factor the double sum into a product of two sums, as shown in the displayed formula. The first factor is $\zeta(k)$, and that's why it factors out and can appear at the head of the formula for the $q$-expansion of $G_k$.

**EXERCISE**: Can you figure out why the number 2 also can be factored out?

It is very handy to divide (13.3) by the factor $2\zeta(k)$ to obtain new modular forms (just multiples of the old ones) whose $q$-expansions all start with 1. So we define

$$E_k(z) = \frac{1}{2\zeta(k)} G_k(z) = 1 + C_k(q + \sigma_{k-1}(2)q^2 + \sigma_{k-1}(3)q^3 + \cdots),$$

where $C_k = -\frac{2k}{B_k}$ is a constant depending on $k$.
  For example,

$$E_4(z) = 1 + 240(q + (1 + 2^3)q^2 + (1 + 3^3)q^3 + \cdots)$$

$$= 1 + 240(q + 9q^2 + 28q^3 + \cdots)$$

and

$$E_6(z) = 1 - 504(q + (1 + 2^5)q^2 + (1 + 3^5)q^5 + \cdots)$$

$$= 1 - 504(q + 33q^2 + 244q^3 + \cdots).$$

Here you need to know the value of the Bernoulli numbers $B_4 = -\frac{1}{30}$ and $B_6 = \frac{1}{42}$. You can see there are lots of special-looking integers here to play around with.

## 4. Multiplying Modular Forms

We've seen that you can multiply a modular form by a scalar (any complex constant) and get a new modular form of the same weight. Also, you can add together two modular forms of the same weight and get a new modular form of that same weight. A very special thing you can do is *multiply* two modular forms of the same *or different* weights to get a new modular form whose weight is the sum of the two weights you started with.

  You can see why multiplying modular forms is a good idea by looking back at the definition of a modular form of weight $k$. Suppose $f(z)$ and $g(z)$ are two functions from the upper half-plane $H$ to the complex numbers $\mathbf{C}$. The conditions of analyticity and growth are satisfied by $fg$ if they are satisfied by $f$ and $g$. As for the transformation law, if $f$ is a modular form of weight $k_1$, then when it is transformed by $\gamma$, the automorphy factor $(cz + d)^{k_1}$ comes out. Similarly, if $g$ is a modular form of weight $k_2$, then when it is transformed by $\gamma$, the automorphy factor $(cz + d)^{k_2}$ comes out. Now transform $fg$ by $\gamma$. What comes out is $(cz + d)^{k_1}(cz + d)^{k_2} = (cz + d)^{k_1+k_2}$, so the automorphy factor for a modular form of weight $k_1 + k_2$ comes out.

For example, we can multiply $E_4$ and $E_6$ together to get a modular form, call it for the time being $H_{10}$, of weight 10. If we want to, we can do the multiplication on their $q$-expansions. Thus, the $q$-expansion of $H_{10}$ is

$$H_{10}(z) = E_4(z)E_6(z)$$

$$= (1 + 240(q + 9q^2 + 28q^3 + \cdots))$$

$$\times (1 - 504(q + 33q^2 + 244q^3 + \cdots)).$$

Although we are being asked to multiply two infinite series together, which might take us an infinite amount of time, we can multiply as many terms as we wish of the beginning of each series to see how the series for the product begins. Thus,

$$H_{10}(z) = 1 - 264q - 135432q^2 - 5196576q^3 - 69341448q^4$$

$$- 515625264q^5 - \cdots.$$

It turns out that any modular form is a polynomial in $E_4$ and $E_6$. This is a rather surprising theorem. It says that if $f(z)$ is any modular form of weight $k$, then there is some polynomial $F(x, y)$ with complex coefficients with the property that when you substitute $E_4$ for $x$ and $E_6$ for $y$, you get $f$ on the nose:

$$f(z) = F(E_4(z), E_6(z)).$$

Because the weights add when we multiply, you can see that this polynomial $F(x, y)$ may be assumed to be homogeneous of weight $k$, if we give $x$ weight 4 and $y$ weight 6:

$$F(x, y) = \sum_{\substack{i,j \\ 4i+6j=k}} c_{ij}x^i y^j.$$

We will see combinations of modular forms like this in later chapters. We can look at a couple of examples right now.

A very important example is $\Delta$, defined by

$$\Delta = \frac{1}{1728}(E_4^3 - E_6^2).$$

This function first arose in the theory of elliptic curves,[5] where it is called the *discriminant*, and we will see it when we study the number of ways an integer can be written as a sum of 24 squares! The weight of $\Delta$ is $12 = 3 \cdot 4 = 2 \cdot 6$. (The expression $3 \cdot 4$ comes from looking at the cube of $E_4$, and similarly the factor $2 \cdot 6$ comes from looking at the square of $E_6$.)

Let's look at the $q$-expansion of $\Delta$. The $q$-expansion of $E_4$ starts with $1 + 240q$. So the $q$-expansion of $E_4^3$ starts with $1 + 3 \cdot 240q$. The $q$-expansion of $E_6$ starts with $1 - 504q$. So the $q$-expansion of $E_6^2$ starts with $1 - 2 \cdot 504q$. Because $3 \cdot 240 = 720$ and $2 \cdot 504 = 1008$, the $q$-expansion of $\Delta$ begins with $(1 - 1) + \frac{1}{1728}(720 - (-1008))q = q$. In summary,

$$\Delta = q + \tau(2)q^2 + \tau(3)q^3 + \cdots.$$

Here we have used Ramanujan's notation $\tau(n)$ for the coefficients beyond the first. It turns out that every value $\tau(n)$, which *a priori* is some rational number, is actually an ordinary integer. The first few values of $\tau(n)$ are $\tau(2) = -24$, $\tau(3) = 252$, $\tau(4) = -1472$, $\tau(5) = 4830$, and $\tau(6) = -6048$. Note that $\tau(2)\tau(3) = \tau(6)$.

We can make three observations:

(1) The reason we divided by 1728 in the definition of $\Delta$ is to make the coefficient of $q$ as simple as possible.

(2) Because $1728 = 2^6 \cdot 3^3$ is the denominator in our definition of $\Delta$, if we look at things modulo 2 or 3 we might expect complications. This is indeed the case: The theory of elliptic curves modulo 2 or 3 is considerably more complicated than modulo other primes. (By the way, it is a little surprising that 1728, which arises as a *sum* of

---

[5] You could refer to Ash and Gross (2012) for an elementary treatment of elliptic curves.

$3 \cdot 240$ and $2 \cdot 504$, should have a prime factorization involving only 2's and 3's. That fact is somehow forced from the theory of elliptic curves and its relationship to $\Delta$.)

(3) The constant term in the $q$-expansion of $\Delta$ is 0. This means that $\Delta$ is a cusp form. It is not easy to write down cusp forms by knowing in advance the coefficients of their $q$-expansions. For instance, there is a lot that is unknown about the values of the $\tau$-function. Indeed, it is not even known whether $\tau(n)$ can ever equal 0. But we can create cusp forms with judiciously chosen polynomial combinations of Eisenstein series. We can choose the coefficients of our polynomial combination so the constant terms cancel out. This is what we did to create $\Delta$, using the lowest possible weight. The powers of $E_4$ have weights $4, 8, 12, \ldots$ and the powers of $E_6$ have weights $6, 12, \ldots$, so the first agreement of weights we get is at 12. In fact, there are no nonzero cusp forms of weight less than 12, and $\Delta$ is the only cusp form of weight 12 (up to scalar multiplication).

Another consequence of the fact that any modular form for $\mathrm{SL}_2(\mathbf{Z})$ is a polynomial in $E_4$ and $E_6$ is that the other Eisenstein series must be polynomials in $E_4$ and $E_6$. For fun, let's look at an example of this. The product $E_4 E_6$ has weight 10, and this is the only way to get a product of Eisenstein series of weight 10. Therefore $E_{10}$, which also has weight 10, must be a multiple of $E_4 E_6$. Since both $E_{10}$ and $E_4 E_6$ have $q$-expansions beginning with 1, this multiple must be unity. In other words, $E_{10}$ is what we called $H_{10}$ earlier:

$$E_{10} = E_4 E_6.$$

Now this formula has a curious consequence for the $\sigma$-functions. We have

$$E_4(z)E_6(z) = \left(1 + \sum_{n=1}^{\infty} 240\sigma_3(n)q^n\right)\left(1 - \sum_{n=1}^{\infty} 504\sigma_5(n)q^n\right)$$

and

$$E_{10}(z) = 1 - \frac{20}{B_{10}} \sum_{n=1}^{\infty} \sigma_9(n)q^n = 1 - \sum_{n=1}^{\infty} 264\sigma_9(n)q^n$$

because $B_{10} = \frac{5}{66}$. If we multiply the first two series out to however many powers of $q$ we like, and equate the coefficients with the latter series, we get interesting identities between the $\sigma$-functions.

To be concrete, we have

$$(1 + 240q + 240\sigma_3(2)q^2 + \cdots)(1 - 504q - 504\sigma_5(2)q^2 + \cdots)$$

$$= 1 - 264q - 264\sigma_9(2)q^2 + \cdots.$$

The constant terms on both sides are equal to 1, as they should be. Equating the coefficients of $q$ on both sides gives $240 - 504 = -264$. Check. Now equate the coefficients of $q^2$:

$$-504\sigma_5(2) - (240)(504) + 240\sigma_3(2) = -264\sigma_9(2).$$

That's a strange formula relating the third, fifth, and ninth powers of the divisors of the number 2. Let's check this formula numerically. The left-hand side is

$$-504(1 + 2^5) - (240)(504) + 240(1 + 2^3)$$

$$= -504 \cdot 33 - 240 \cdot 504 + 240 \cdot 9 = -135432.$$

On the right-hand side, we have

$$-264(1 + 2^9) = -135432.$$

If you like arithmetic, you can equate the coefficients of $q^3$ and see what you get.

## 5. Dimensions of $M_k$ and $S_k$

As you can see so far (and we'll have other examples in the next chapter), it's very handy to know first that $M_k$ and $S_k$ are finite-dimensional vector spaces and second what their dimensions are.

How do we know this? We can't go into any details here, but we can say something vague.

First, you take the fundamental domain $\Omega$ from figure 12.1. Any modular form is determined by its values on $\Omega$. Now $\Omega$ is much smaller than the whole upper half-plane $H$. It's a bit lopsided though. We have included the left-hand portion of its boundary but not its right-hand portion. It is much more equitable to work with $\overline{\Omega}$, where we include both boundaries. (This is called the *closure* of $\Omega$.) But $\overline{\Omega}$ is a little too big to be a fundamental domain. If $z$ is a point on its right-hand border, then the point $z - 1$ is on its left-hand border, and the two points are in the same orbit of $SL_2(\mathbf{Z})$. (In fact, $z = \gamma(z - 1)$, where $\gamma = \left[\begin{smallmatrix} 1 & 1 \\ 0 & 1 \end{smallmatrix}\right]$.) Also, a point $z$ on the right half of the semicircle is in the same orbit as a certain point on the left half, namely $-1/z$. (Here the matrix relating them is $\left[\begin{smallmatrix} 0 & 1 \\ -1 & 0 \end{smallmatrix}\right]$.)

So the fair thing to do is to work with all of $\overline{\Omega}$ but to "identify" or "sew together" (à la topology) the right and left vertical borders and the right and left semicircles by attaching each $z$ in the boundary of $\overline{\Omega}$ to the other point in its orbit in the boundary.

When we do this sewing, we get something that looks like a stocking with a very pointy toe at $\rho$. (The point $\rho$ is the sixth root of unity on the right.) There is also a less pointy place in the heel at $i$ (the square root of $-1$). Other than these two "singular" points, the rest of the stocking is nice and smooth. Because we built this shape out of a piece of the complex plane, the stocking still is a "complex space," meaning we can do complex analysis on it. Let's call this stocking $Y$.

There is a way to smooth out the two singular points $\rho$ and $i$ to make all of $Y$ into what is called a *Riemann surface*. The Riemann surface is still unsatisfactory because it is not "compact." This problem means that you can go along the shank of the stocking infinitely far forever. (This corresponds to letting the imaginary part of $z$ go to $\infty$ in $\overline{\Omega}$.) We can resolve the compactness issue by sewing up the mouth of the stocking, and when we sew it up in the right way (which corresponds to taking the $q$-expansion seriously as a Taylor series), then we get a compact Riemann surface. It looks something like a very misshapen sphere. Topologically, it is a sphere, but it is a sphere that remembers it had some funny points on it, namely those coming from $\rho$, $i$, and the cusp at "infinity." We call this sphere $X$. So $X = Y +$ a point at infinity.[6]

Now, a compact Riemann surface is a really great mathematical object. We can do complex analysis on it—differentiate, integrate, and so on. We can reinterpret the concept of modular form to view it as a kind of function on $X$. It is not really a function, because the automorphy factor gets in the way, but it is a *section of a vector bundle*—whatever that means. The point is that in such a context the space of sections is a finite-dimensional vector space, and something called the Riemann–Roch Theorem lets us compute the dimension of this vector space. Just think of this as using the tools of complex analysis, such as the residue theorem, in a sophisticated way. In fact, you can phrase everything in terms of residues if you want.

To summarize, once we have rethought a modular form as some kind of analytic object that lives on a complex Riemann surface, it follows automatically that the modular forms of a given positive integral weight $k$ make up a finite-dimensional vector space. Moreover, we can compute the dimension by remembering the funny points $\rho$, $i$, and the cusp at "infinity" and using complex analysis.

What are the dimensions? Here is the answer. Let $m_k$ denote the dimension of the space of modular forms $M_k$, and $s_k$ the dimension of the space of cusp forms $S_k$. We have already seen that whatever $k$ is (provided that $k$ is a nonnegative even integer), $s_k = m_k$ or $s_k = m_k - 1$. (As usual in this chapter, we are looking at modular forms for the full modular group $SL_2(\mathbf{Z})$.)

First of all, if $k = 0$, there are the constant functions. They are obviously in $M_0$. (Check the definitions.) They make up a vector space of dimension 1, with basis given by a single element: the constant function 1. But there are no cusp forms of weight 0 other than the 0 function, so $m_0 = 1$ and $s_0 = 0$.

---

[6] If you work with a congruence subgroup $\Gamma$ of $SL_2(\mathbf{Z})$, then you can do similar things. There is a nice fundamental domain. You can take its closure, smooth out its singular points, add in its cusps, and get a compact Riemann surface, which is called the *modular curve* corresponding to $\Gamma$. It is called a curve rather than a surface because in number theory we are interested in its properties as an algebraic object over $\mathbf{C}$, from which point of view it has dimension 1.

TABLE 13.1. Values of $m_k$ and $s_k$

| k | 0 | 2 | 4 | 6 | 8 | 10 | 12 | 14 | 16 | 18 | 20 | 22 | 24 | 26 | 28 | 30 | 32 | 34 | 36 | 38 | 40 |
|---|---|---|---|---|---|----|----|----|----|----|----|----|----|----|----|----|----|----|----|----|----|
| $m_k$ | 1 | 0 | 1 | 1 | 1 | 1 | 1 | 2 | 1 | 2 | 2 | 2 | 2 | 3 | 2 | 3 | 3 | 3 | 3 | 4 | 3 | 4 |
| $s_k$ | 0 | 0 | 0 | 0 | 0 | 0 | 1 | 0 | 1 | 1 | 1 | 1 | 2 | 1 | 2 | 2 | 2 | 2 | 3 | 2 | 3 |

If the weight $k = 2$, it turns out that $m_2 = s_2 = 0$. There are no modular forms of weight 2, or, to be more accurate, the only modular form of weight 2 is the 0 function. (The 0 function is a modular form and a cusp form of any weight.)

For any weight $k \geq 4$, it is true that $m_k = s_k + 1$, because there is an Eisenstein series of each weight $4, 6, 8, 10, \ldots$.

If $k = 4, 6, 8,$ or $10$, then the only modular forms of weight $k$ are scalar multiples of Eisenstein series, so $m_k = 1$ and $s_k = 0$ for those weights.

Starting with $k = 12$ and going on forever, the dimensions $m_k$ and $s_k$ have a sort of periodic behavior with period 12. Every time you move up by 12, you add 1: $m_{k+12} = m_k + 1$ and $s_{k+12} = s_k + 1$. To begin with, we have

$$m_{12} = 2 \quad m_{14} = 1 \quad m_{16} = 2 \quad m_{18} = 2 \quad m_{20} = 2 \quad m_{22} = 2.$$

To get the next batch, just add one:

$$m_{24} = 3 \quad m_{26} = 2 \quad m_{28} = 3 \quad m_{30} = 3 \quad m_{32} = 3 \quad m_{34} = 3,$$

and so on. For $k \geq 4$, if you want to know $s_k$, just subtract 1 from $m_k$. For example, $s_{12} = 1$, in line with our assertion that the only cusp forms of weight 12 are scalar multiples of $\Delta$. We will see what this means for sums of 24 squares in chapter 15.

We can summarize this periodicity in a strange-looking two-part formula for $m_k$, valid for $k > 0$,

$$m_k = \begin{cases} \left\lfloor \dfrac{k}{12} \right\rfloor + 1 & k \not\equiv 2 \pmod{12} \\[2ex] \left\lfloor \dfrac{k}{12} \right\rfloor & k \equiv 2 \pmod{12}, \end{cases}$$

where the notation $\lfloor x \rfloor$ refers to the largest integer no larger than $x$. The first few values of $m_k$ and $s_k$ are listed in table 13.1. The funny dip in the dimensions for weight $k = 14, 26, \ldots$ comes out of the computations with the Riemann–Roch formula.

Chapter 14

# CONGRUENCE GROUPS

## 1. Other Weights

The main ideas of modular forms are captured by what we have said about modular forms for the whole group $\mathrm{SL}_2(\mathbf{Z})$. You can just imagine that similar things hold for modular forms for congruence subgroups, which we will define right away. However, to discuss the applications of modular forms we have in mind for the following chapters, we will have to refer to these congruence groups.

From an abstract, noodling-around point of view, we could be led to the congruence groups by asking about modular forms $f$ of weight $k$, when $k$ is not a positive even integer. We will always want a modular form to be an analytic function on the upper half-plane $H$. Also, we will always require some kind of growth condition on $f$ near the cusps, which from now on we will assume without specifying carefully. The growth condition is important when you work out the theory in detail, but to get a feel for the basic ideas, we can leave it in the background.

The main thing is the transformation law for $f$ under matrices $\gamma$ in $\mathrm{SL}_2(\mathbf{Z})$. Remember that if $f$ is to have weight $k$, then we require

$$f(\gamma(z)) = (cz + d)^k f(z)$$

for all $\gamma = \begin{bmatrix} a & b \\ c & d \end{bmatrix}$ in $\mathrm{SL}_2(\mathbf{Z})$, where $\gamma(z) = \frac{az+b}{cz+d}$.

To begin with, let's wonder what happens to modular forms of odd integral weight. Suppose $f(z)$ is a modular form of weight $k$ for all of $\mathrm{SL}_2(\mathbf{Z})$, where we assume $k$ is positive and an odd integer. Consider the transformation law for the matrix $\gamma = -I$, where $-I$

is the 2-by-2 matrix

$$-I = \begin{bmatrix} -1 & 0 \\ 0 & -1 \end{bmatrix}$$

Because $\gamma(z) = z$ and $cz + d = -1$ for this $\gamma$, the transformation law for $\gamma$ gives $f(z) = (-1)^k f(z) = -f(z)$. This implies that $f(z)$ must be identically 0. We say there are "no" modular forms of odd weight for the full modular group. (Of course, the constant function 0 is a modular form of *any* weight, so take this "no" with a grain of salt. "No" means "none except for the 0 function.")

If (even worse) we let the weight $k$ be a fraction $\frac{p}{q}$, where $p$ and $q$ are relatively prime and $q \neq 1$, then we may begin to wonder about which $q$th root to take when interpreting the expression $(cz + d)^{p/q}$. (There are other things to worry about, too, but we will get to them later.)

We can get modular forms of other weights if we relax the transformation law. We may simply declare that the transformation law doesn't have to be valid for every $\gamma$ in $\mathrm{SL}_2(\mathbf{Z})$ but only for some $\gamma$'s.

For instance, let's go back to odd integral weights. If we impose some congruence conditions on $a$, $b$, $c$, and $d$ in the allowable $\gamma$'s, we can eliminate $-I$ from our group. For example, if we define $\Gamma_1(3)$ to be the set of matrices $\begin{bmatrix} a & b \\ c & d \end{bmatrix}$ in $\mathrm{SL}_2(\mathbf{Z})$ such that $a \equiv d \equiv 1$ (mod 3) and $c \equiv 0$ (mod 3), then $-I$ is not in $\Gamma_1(3)$. So if we require the transformation law only for $\gamma$'s in *this* set of matrices $\Gamma_1(3)$, we do not have to consider the fatal transformation under $-I$. In fact, there are plenty of modular forms of odd weight for $\Gamma_1(3)$.

It turns out that the theory of modular forms of fractional weights also works better with congruence subgroups. So let's make some precise definitions.

**DEFINITION**: A *subgroup* of $\mathrm{SL}_2(\mathbf{Z})$ is a subset $K$ of $\mathrm{SL}_2(\mathbf{Z})$ with the properties

(1) $I$ is in $K$.
(2) If $A$ is in $K$, so is $A^{-1}$.
(3) If $A$ and $B$ are in $K$, then so is $AB$.

One more bit of terminology: If $K$ is a subgroup of $SL_2(\mathbf{Z})$ and $L$ is a subset of $K$ that is also a subgroup of $SL_2(\mathbf{Z})$, then we sometimes say for short that $L$ is a subgroup of $K$.

For some examples of subgroups of $SL_2(\mathbf{Z})$—ones that we will be referring to repeatedly—we start by picking a positive integer $N$. Define

$$\Gamma(N) = \left\{\gamma \in SL_2(\mathbf{Z}) \mid \gamma \equiv I \pmod{N}\right\}.$$

In words, $\Gamma(N)$ consists of all the matrices $\left[\begin{smallmatrix} a & c \\ b & d \end{smallmatrix}\right]$ where $a$, $b$, $c$, and $d$ are integers, $ad - bc = 1$, and $a - 1$, $b$, $c$, and $d - 1$ are all divisible by $N$. (Notice that $\Gamma(1)$ is in fact $SL_2(\mathbf{Z})$.) We leave it to you to check that $\Gamma(N)$ is truly a subgroup of $SL_2(\mathbf{Z})$. That means that you must check that the identity matrix $I$ is in $\Gamma(N)$, that the product of two matrices in $\Gamma(N)$ is still in $\Gamma(N)$, and that the inverse of a matrix in $\Gamma(N)$ is in $\Gamma(N)$.

There are two other important families of subgroups of $SL_2(\mathbf{Z})$:

$$\Gamma_0(N) = \left\{\gamma = \begin{bmatrix} a & b \\ c & d \end{bmatrix} \in SL_2(\mathbf{Z}) \;\middle|\; c \equiv 0 \pmod{N}\right\},$$

$$\Gamma_1(N) = \left\{\gamma = \begin{bmatrix} a & b \\ c & d \end{bmatrix} \in SL_2(\mathbf{Z}) \;\middle|\; c \equiv a - 1 \equiv d - 1 \equiv 0 \pmod{N}\right\}.$$

Again you can check that they are subgroups using the definition.

These definitions may seem innocuous, but they pack a punch. They give us some of the most potent subgroups of $SL_2(\mathbf{Z})$ for number-theoretical purposes, especially for studying modular forms. Because they are defined using congruences, they are examples of "congruence groups." Notice that $\Gamma(N)$ is a subgroup of $\Gamma_1(N)$, and in turn $\Gamma_1(N)$ is a subgroup of $\Gamma_0(N)$. Also, if $N|M$, then $\Gamma(M)$ is a subgroup of $\Gamma(N)$, $\Gamma_0(M)$ is a subgroup of $\Gamma_0(N)$, and $\Gamma_1(M)$ is a subgroup of $\Gamma_1(N)$.

The most general congruence subgroup of $SL_2(\mathbf{Z})$ is easy to define.

**DEFINITION**: $K$ is a *congruence group* if

(1) $K$ is a subgroup of $SL_2(\mathbf{Z})$.
(2) For some positive integer $N$, $\Gamma(N)$ is a subgroup of $K$.

If in condition (2) you take the *smallest* $N$ for which $\Gamma(N)$ lies inside $K$, then $N$ is called the *level* of $K$. For example, the levels of $\Gamma(N)$, $\Gamma_0(N)$, and $\Gamma_1(N)$ are all $N$. Note that $\mathrm{SL}_2(\mathbf{Z})$ itself has level 1 and is the only congruence group of level 1.

## 2. Modular Forms of Integral Weight and Higher Level

We can extend our definitions for modular forms to any congruence group $\Gamma$. Let $k$ be an integer. We can now get nonzero examples of modular forms even if $k$ is an odd positive integer. We say that $f(z)$ is a modular form of weight $k$ for $\Gamma$ if

(1) $f(z)$ is an analytic function on the upper half-plane $H$.
(2) $f(z)$ satisfies some growth condition at the cusps that we won't specify.
(3) $f(z)$ satisfies the usual transformation law for matrices in $\Gamma$. Precisely,
$$f(\gamma(z)) = (cz + d)^k f(z)$$
for all $\gamma$ in $\Gamma$, where as usual $\gamma = \left[\begin{smallmatrix} a & b \\ c & d \end{smallmatrix}\right]$. Note that we might get lucky and $f$ might satisfy the transformation law for some other matrices as well. In other words, putting this backward, if $f$ is a modular form for $\Gamma$, then it will also be considered a modular form for every congruence subgroup of $\Gamma$.
(4) $f(z)$ is *analytic* at all the cusps. In fact, this will imply condition (2).
(5) We call $f(z)$ a *cusp form* if it vanishes at all the cusps.

If $\Gamma$ has level $N$, you might think we could say $f(z)$ has level $N$. For technical reasons, we do not say this. However, if $\Gamma = \Gamma_1(N)$, then we do say $f$ has level $N$.

To explain (2), (4), and (5), we have to start a new section.

## 3. Fundamental Domains and Cusps

Remember from section 3 in chapter 12 that $\mathrm{SL}_2(\mathbf{Z})$ had a fundamental domain $\Omega$ for how it "acted" on the upper half-plane $H$.

Namely, for every point $z$ in $H$, there was some point $z_0$ in $\Omega$ and some $\gamma$ in $SL_2(\mathbf{Z})$ with the property that $\gamma(z_0) = z$. You can think of $z_0$ as the "home base" for the orbit of $SL_2(\mathbf{Z})$ through $z$. The other condition that $\Omega$ has to satisfy was that if $z_0$ and $z_1$ are in $\Omega$ and $\gamma$ is in $SL_2(\mathbf{Z})$, and if $\gamma(z_0) = z_1$, then $z_0 = z_1$. In other words, there should be no redundancy in $\Omega$: It contains one and only one point from each orbit. It follows that a modular form $f(z)$ has all its values determined just from its values on $\Omega$.

Now suppose that $f(z)$ is a modular form for a fixed congruence group $\Gamma$. We only know the transformation law for $\gamma$ in $\Gamma$, which may well be considerably smaller than $SL_2(\mathbf{Z})$. From the values of $f$ on $\Omega$, we don't have enough information to determine the values of $f$ everywhere. What should we do?

The obvious thing is to enlarge $\Omega$ so that it becomes a fundamental domain for $\Gamma$. It turns out that there is a *finite* set of matrices $g_1, \ldots, g_t$ in $SL_2(\mathbf{Z})$ with the following property:

If we define

$$\Omega_i = g_i\Omega = \left\{ z \in H \mid z = g_i(w) \text{ for some } w \in \Omega \right\},$$

and then we form the union $\Psi = \Omega_1 \cup \Omega_2 \cup \cdots \cup \Omega_t$, then $\Psi$ is a fundamental domain for $\Gamma$ in exactly the same sense that $\Omega$ is a fundamental domain for $SL_2(\mathbf{Z})$.

For example, figure 14.1 contains a diagram of a fundamental domain for $\Gamma_0(3)$ obtained by using the standard fundamental domain in figure 12.1 as $\Omega$.

There will be a finite number of points on $\mathbf{R} \cup \{i\infty\}$ to which $\Psi$ will "go out" in the same way that $\Omega$ "goes out" to $i\infty$. These points are the *cusps* of $\Psi$. The fundamental domain $\Psi$ is not unique, but the orbits of its cusps are determined only by $\Gamma$ and not by any of the other choices that we have made. Altogether, $\Psi$ will have a certain number of cusps. They are not actually points in $\Psi$ or in $H$ but rather points on the real line or the "point" at $i\infty$. For example, any fundamental domain for $\Gamma_0(3)$ has exactly two cusps.

Every one of these cusps behaves like "the" cusp for $\Omega$. Remember that a modular form for $SL_2(\mathbf{Z})$ is periodic of period 1 and therefore

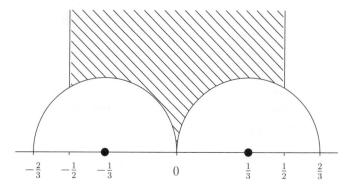

Figure 14.1. Fundamental domain for $\Gamma_0(3)$

has a $q$-expansion. One of our conditions for $f(z)$ to be a modular form was that its $q$-expansion have no negative powers of $q$, and the further condition for $f(z)$ to be a cusp form was for the constant term of its $q$-expansion to be 0.

It turns out that a similar thing is true for each cusp of $\Psi$ given a modular form for $\Gamma$ (where $\Psi$ is a fundamental domain for $\Gamma$). At each of these cusps, there is a $\tilde{q}$-expansion, and condition (4) in the previous section means that this $\tilde{q}$-expansion has no negative powers of $\tilde{q}$ for each and every cusp. Of course, this $\tilde{q}$ depends on the cusp. It is not $e^{2\pi i z}$ but a similar function. The meaning of "vanishes" at all the cusps should now be clear: It means that the $\tilde{q}$-expansion has constant term 0 at each and every cusp.

## 4. Modular Forms of Half-Integral Weight

We can now let the weight $k$ be fractional. For our purposes in this book, we need only consider the case when $k$ is a half-integer. This means that $k = t/2$, where $t$ is an odd integer. To quote Buzzard (2013), "The theory for weight $\mathbf{Z} + \frac{1}{2}$ is sufficiently different from that of weight in $\mathbf{Z}$ that we specifically exclude weight [in] $\mathbf{Z}$ [when defining modular forms of half-integral weight]." Although the theory of half-integral–weight modular forms is rich and fascinating, it is very complicated, and we prefer to avoid it in this book. However, because we wanted to use the modular form $\eta$, which is a form of weight $\frac{1}{2}$, we have included this very short section. Additionally, consideration of sums of an odd number of squares leads naturally to half-integral–weight forms.

There are different ways to define modular forms of half-integral weight. To get the transformation laws to work, we would have to do some fancy footwork involving what is called a "multiplier system." So that is one way to make a definition: Copy the definition for integral $k$ but allow multiplier systems. (You also need to be careful with the growth condition, but we won't worry about that here.)

The theory gets simpler if we give up on level 1. Remember that level 1 means we require a nice transformation law for every matrix in $\text{SL}_2(\mathbf{Z})$. A simpler theory can be developed if we stick to smaller congruence groups. In particular, suppose that we require $\Gamma$ to be a congruence subgroup of $\Gamma_0(4)$. Then a modular form of half-integral weight $k$ is an analytic function on $H$ that transforms by

$$f(\gamma(z)) = j(\gamma, z)f(z)$$

for each $\gamma$ in $\Gamma$. Here $j(\gamma, z)$ is a function related to $(cz + d)^k$ but quite a bit more complicated. For more information, you could read Buzzard (2013) or look at the references in that paper.

As usual, we require $f$ to be analytic at the cusps. Once again, the space of modular forms of level $N$ and half-integral weight $k$ is a finite-dimensional vector space. Quite a bit is known about this space, but the theory is well beyond the scope of this book.

# Chapter 15

# PARTITIONS AND SUMS OF SQUARES REVISITED

## 1. Partitions

In this chapter, we discuss a couple of examples of how we can use modular forms to solve some of the problems posed earlier in the book. The proofs of most of these results are too advanced to include, but we can give glimpses of why modular forms come into the game.

Recall that if $p(n)$ is the number of partitions of $n$ into positive parts (not respecting the order of the parts), then we know its generating function looks like this:

$$F(q) = \sum_{n=1}^{\infty} p(n)q^n = \prod_{i=1}^{\infty} \frac{1}{1-q^i}.$$

We've switched the variable in the generating function from $x$ to $q$ because we are going to set $q = e^{2\pi i z}$ and view $F(q)$ as a function on the upper half-plane.

It is often true that if we have a nontrivial identity between an infinite sum and an infinite product, then we have something

(1) interesting,
(2) useful, and
(3) hard to prove.

This particular identity is not hard to prove, but it is interesting and is very useful for studying the partition function.

Because we made $F$ a function of $q$, we automatically know that $F$ when viewed as a function of $z$ is periodic of period 1. Could $F(q)$ be a modular form? Well, no. But it can be related to modular forms, as we will see later in this section. Exploiting the relationship, one can prove some very nice theorems. (The results we are going to mention can also be proven by other methods. There are other, more abstruse results whose proof—as far as anyone knows—requires the theory of modular forms and related methods.) We will mention two theorems, one on the size of $p(n)$ and one on congruences satisfied by $p(n)$.

How big is $p(n)$? If you do a few computations by hand, you will quickly get tired, because $p(n)$ gets big rather rapidly. But how big? One way to answer this question is to give an explicit formula for $p(n)$ in terms of other functions of $n$ and then see how big the other functions get. Although there are exact formulas for $p(n)$ in terms of other functions of $n$, the functions involved are rather complicated. A more perspicuous answer is given by asking for an *approximate* formula for $p(n)$ in terms of other functions of $n$ whose sizes are easy to understand.

For instance, suppose you had two functions of $n$, say $f(n)$ and $g(n)$. Think of $f$ as "unknown" and $g$ as friendly. We use the notation

$$f(n) \sim g(n)$$

to assert that

$$\lim_{n \to \infty} \frac{f(n)}{g(n)} = 1.$$

Such a formula is called an *asymptotic formula* for $f$ in terms of $g$. For example, the Prime Number Theorem gives an asymptotic formula for the function $\pi(N)$.

The following asymptotic formula for the size of $p(n)$ is due to Hardy and Ramanujan:

$$p(n) \sim \frac{e^{a(n)}}{b(n)},$$

where $a(n) = \pi\sqrt{2n/3}$ and $b(n) = 4n\sqrt{3}$. The asymptotic formula shows that $p(n)$ grows more slowly than $e^n$ but faster than any polynomial in $n$.

Another kind of result about $p(n)$ that is proved by using modular forms is congruences, due to Ramanujan, concerning certain values of $p(n)$ modulo 5, 7, and 11:

$$p(5k + 4) \equiv 0 \pmod{5},$$

$$p(7k + 5) \equiv 0 \pmod{7},$$

$$p(11k + 6) \equiv 0 \pmod{11}.$$

These are congruences for $p(n)$ with modulus a prime $m$, with $n$ in an arithmetic progression of the form $n = mk + d$, when $m = 5$, 7, or 11. Ahlgren and Boylan proved, among other things, that there are no such congruences modulo any other prime, while for 5, 7, and 11 you only get them for $5k + 4$, $7k + 5$, and $11k + 6$.[1] The fact that the primes involved here are the primes less than 12 but not equal to 2 or 3 is related in some mysterious way to the fact that $\Delta$ has weight 12. Primes dividing 12 are "special."[2]

What relationship can we find between $F(q)$ and modular forms? One such relationship comes from a remarkable formula for the cusp form of weight 12 we called $\Delta$. Remember from section 4 of chapter 13 that we defined

$$\Delta = \frac{1}{1728}(E_4^3 - E_6^2) = \sum_{n=1}^{\infty} \tau(n)q^n.$$

The coefficients $\tau(n)$ are not understood perfectly. We follow Ramanujan and just name them $\tau(n)$. Of course, we can compute them for "small" $n$ using a computer.

Jacobi proved the amazing formula

$$\Delta = q \prod_{i=1}^{\infty} (1 - q^i)^{24}.$$

There's that pesky 12 showing up again, except it doubled itself to 24.

We can manipulate Jacobi's formula into the generating function $F(q)$ for the partition function by dividing by $q$ and taking the 24th

---

[1] Many *other* kinds of congruences for the partition function have been proven.

[2] There do not seem to be any known nontrivial general congruences modulo 2 or 3. In particular, no simple formula is known that could be used, given any $n$, to tell whether $p(n)$ is even or odd.

root and then the reciprocal. This gives

$$(q/\Delta)^{1/24} = \prod_{i=1}^{\infty}(1 - q^i)^{-1} = F(q).$$

That was easy! The big problem here is taking the 24th root. Because we are dealing with complex-valued functions of a complex variable, there are 24 possible 24th roots for each nonzero complex value, and there is no reason to believe that we can choose them in a coherent way. A smaller problem is that the reciprocal of a modular form is usually not a modular form. Also, $q$ by itself is not a modular form, so it is not $F(q)$ but $F(q)/q^{1/24}$ that should have something to do with modular forms. But we do see that $F(q)$ might be related to something of weight $\frac{1}{2}$. Why? The function $\Delta$ has weight 12, and taking its 24th root (if we could do that nicely) would give something of weight $\frac{12}{24} = \frac{1}{2}$.

Dedekind defined the $\eta$-function by the formula

$$\eta(q) = q^{1/24}\prod_{i=1}^{\infty}(1 - q^i),$$

where we define $q^{1/24} = e^{i\pi z/12}$. (Notice that this choice of 24th root is unambiguous.) With this notation, we see that the generating function of the partition function satisfies the equation

$$\frac{q^{1/24}}{F(q)} = \eta(q).$$

Thus, things we can discover about $\eta(q)$ can easily imply new facts about $F(q)$ and hence about $p(n)$.

This function $\eta(q)$ *ought* to be a modular form of weight $\frac{1}{2}$. Of course, we have to be careful to say what we mean by a modular form of weight $\frac{1}{2}$. In the transformation law, we first have to worry about which square root of $(cz + d)$ to take, but it turns out that it is not hard to make a consistent choice. What's more serious is that if we want a nice transformation law for every $\gamma$ in $\mathrm{SL}_2(\mathbf{Z})$, we need to multiply what we might have thought was the correct automorphy factor by a very complicated root of unity, depending on $\gamma$. The rule that tells what root of unity to use is called a "multiplier system." Modifying our definitions appropriately, we could call $\eta$ a "modular form of weight $\frac{1}{2}$ with multiplier system so-and-so" and develop the properties of such forms.

Note that $\eta^{24} = \Delta$. Because $\Delta$ is a fine modular form of even weight, it doesn't need a multiplier system. We can conclude that the multipliers for $\eta$ must be 24th roots of unity.

To summarize, the generating function for the partition function $p(n)$ is closely related to various modular forms. Thus, knowledge of these modular forms and their properties enables one to prove various rather surprising theorems about the number of partitions of $n$, two of which were mentioned at the beginning of this section.

## 2. Sums of Squares

Let's start with

$$\frac{\eta(q)}{q^{1/24}} = \prod_{i=1}^{\infty}(1 - q^i).$$

Euler proved that

$$\prod_{i=1}^{\infty}(1 - q^i) = \sum_{m=-\infty}^{\infty} (-1)^m q^{m(3m-1)/2}.$$

There are many proofs of this identity. Euler found one based on no more than high school algebra, but his argument is very clever. The exponents of $q$ on the right-hand side are called *generalized pentagonal numbers*. They begin 0, 1, 2, 5, 7, 12, 15, 22, 26,... in increasing numerical order. If you take every other entry in the list, starting with 1, you get the usual pentagonal numbers, the fitness of whose name can be seen in figure 5.4.

Now, the left-hand side of Euler's identity is closely related to modular forms, in particular of weight $\frac{1}{2}$. The right-hand side is a sum of terms, each with coefficient $\pm 1$, with exponent a quadratic function of $m$. We can try something even simpler but of the same ilk: Let's make all the coefficients 1 and the exponents simply $m^2$. We obtain a series in $q$ that is traditionally (going back to Jacobi) called $\theta$. So

$$\theta(q) = \sum_{m=-\infty}^{\infty} q^{m^2} = 1 + 2q^2 + 2q^4 + 2q^9 + \cdots,$$

where, as usual, $q = e^{2\pi i z}$.

Because $q$ is a function of $z$, we can view $\eta$ and $\theta$ also as functions of $z$. Our hope is that $\theta(z)$ is related to some modular form or other. But the truth surpasses our hopes (at least this once). In fact, $\theta(z)$ *is* a modular form of weight $\frac{1}{2}$ of level[3] 4. Because we have defined $\theta(z)$ by a power series in $q$, it is automatically periodic of period 1. The main part of the proof, roughly speaking, involves expressing $\theta(-1/z)$ in terms of $\theta(z)$. This can be done using something called the *Poisson summation formula*, which involves Fourier series. It is interesting that this kind of theta function also arises in Fourier's theory of heat conduction.

Modular forms of half-integral weight are rather harder to deal with than forms of integral weight. To fix this, we can look at the square $\theta^2(z)$. It is a bona fide modular form of weight 1 of level 4.

Let's look at the $q$-expansion of $\theta^2$:

$$\theta^2(z) = \left( \sum_{m=-\infty}^{\infty} q^{m^2} \right)^2 = \sum_{n=0}^{\infty} c(n) q^n.$$

How can we interpret the coefficients $c(n)$? Well, how do we get a nonzero term involving $q^n$ when we take the square? We must have two of the $m^2$ that add up to $n$. How many ways can we do that? It's the number of ways of writing $n$ as a sum of two squares, counting[4]

---

[3] See the previous chapter for the concept of level. If you skipped that chapter, you just need to know that this means that $\theta$ doesn't transform so nicely under *all* the matrices in $SL_2(\mathbf{Z})$ but only under a certain subset of them. Because the weight of $\theta$ is a half-integer, the transformation law is somewhat more complicated than what we wrote down for integral weights.

[4] Now you can see why, in chapter 10, we wanted to count the number of ways $n$ was a sum of two squares in this fussy manner. With partitions, we order the parts from smallest to largest, and disordered partitions don't enter into the count. In each problem, the "good" method of counting is determined by the exact relationship of each problem to modular forms. If we used a "bad" method of counting, we wouldn't have any good method of proof until we restored the "good" method of counting.

different orders of the squares in distinct ways and also counting $m$ and $-m$ separately if $m \neq 0$.

Recall that in chapter 10 we defined $r_t(n)$ as the number of ways of writing $n$ as a sum of $t$ squares,

$$n = m_1^2 + m_2^2 + m_3^2 + \cdots + m_t^2,$$

taking into consideration different orders and signs of the integers $m_1, \ldots, m_t$. The same reasoning as in the previous paragraph tells us that

$$\theta^t(z) = \left( \sum_{m=-\infty}^{\infty} q^{m^2} \right)^t = \sum_{n=0}^{\infty} r_t(n) q^n.$$

And we know that $\theta^t(z)$ is a modular form of weight $t/2$ and level 4.

By studying the space of modular forms of weight $t/2$ and level 4 in some other way, we can identify which element of this space is equal to $\theta^t(z)$. In this way, we can prove formulas for $r_t(n)$ for various $t$. Because forms of half-integral weight are more recondite, we now understand why the formulas for $r_t(n)$ with odd $t$ are more complicated and harder to come by than those for even $t$. Also, it turns out that when $t$ is divisible by higher and higher powers of 2 (up to a point), the modular form $\theta^t(z)$ transforms nicely under more matrices than just those in $\Gamma_1(4)$, and so we get even nicer formulas. The best values of $t$ are multiples of 8.

For example, when $t = 2$, we can study the space of modular forms of weight 1 and level 4. This enables us to derive a theorem that says that the number of ways $r_2(n)$ of writing the positive integer $n$ as a sum of two squares (counted in the "good" way) is given by

$$r_2(n) = 4(d_1(n) - d_3(n)),$$

where $d_i(n)$ is the number of positive divisors of $n$ congruent to $i$ (mod 4) for $i = 1$ and 3.

This equation for $r_2(n)$ can also be proved using other methods. As we mentioned in section 2 of chapter 10, an elementary proof can be given using unique factorization of the Gaussian integers $\{a + bi \mid a, b \in \mathbf{Z}\}$. Another way to prove it is to derive a certain formula for the square of the $q$-expansion of $\theta$, obtained from a theory called "the theory of elliptic functions," which is how Jacobi did it. In fact, in this way, Jacobi found and proved similar formulas for the number of ways of writing $n$ as a sum of four, six, and eight squares.

But if we want to understand how to write $n$ as a sum of a large number of squares, the best way of looking at this problem is through the lens of the theory of modular forms. In this way, we can get formulas for $r_t(n)$ in terms of the coefficients of the $q$-expansions of modular forms. As an example of this, we will work out the case for sums of 24 squares, which was first done by Ramanujan. (There are similar formulas, of varying complexity, for sums of other numbers of squares.) A very nice explanation of this may be found in Hardy (1959, pp. 153–57). Hardy gives much more detail of how the proof goes than we are able to do here. (Beware that Hardy uses an older notation that differs from what we use in this book.)

First of all, it turns out that $\theta^{24}(z)$ is modular for a certain congruence group $\Gamma$ that is larger than $\Gamma_1(4)$. This makes it rather easy to identify $\theta^{24}(z)$ among the space of modular forms of weight 12 for $\Gamma$. How? Well, we know a modular form of weight 12 and level 1, namely $\Delta(z)$. We can get other modular forms of weight 12 by replacing $z$ with a nontrivial rational multiple of $z$. The resulting function is no longer modular for the full group $\mathrm{SL}_2(\mathbf{Z})$, but you can check that it will be modular for a smaller congruence group. That is, it will have higher level.

For example, consider the function of $z$ given by $\Delta(2z)$. It won't transform properly under general matrices $\left[\begin{smallmatrix} a & b \\ c & d \end{smallmatrix}\right]$ in $\mathrm{SL}_2(\mathbf{Z})$. But suppose $\gamma = \left[\begin{smallmatrix} a & b \\ c & d \end{smallmatrix}\right]$ is in $\Gamma_0(2)$, meaning that we restrict $c$ to be even. In this case, $\left[\begin{smallmatrix} a & 2b \\ \frac{c}{2} & d \end{smallmatrix}\right]$ is also in $\mathrm{SL}_2(\mathbf{Z})$. Therefore

$$\Delta\left(\frac{az + 2b}{\frac{c}{2}z + d}\right) = \left(\frac{c}{2}z + d\right)^{12} \Delta(z).$$

Replacing $z$ by $2z$ gives

$$\Delta\left(\frac{a(2z) + 2b}{\frac{c}{2}(2z) + d}\right) = \left(\frac{c}{2}(2z) + d\right)^{12} \Delta(2z) = (cz + d)^{12}\Delta(2z).$$

But

$$\frac{a(2z) + 2b}{\frac{c}{2}(2z) + d} = 2\frac{az + b}{cz + d}.$$

We conclude that

$$\Delta(2\gamma(z)) = \Delta\left(2\frac{az + b}{cz + d}\right) = (cz + d)^{12}\Delta(2z),$$

and this is the correct transformation for $\Delta(2z)$ under $z \to \gamma(z)$.

Therefore, $\Delta(2z)$ is a modular form of weight 12 for $\Gamma_0(2)$. When we look at the $q$-expansion of this form, remember that $q = q(z) = e^{2\pi iz}$. Therefore $q(2z) = e^{2\pi i(2z)} = e^{2(2\pi iz)} = q^2$. We defined the $\tau$-function by

$$\Delta(z) = \sum_{n=1}^{\infty} \tau(n)q(z)^n.$$

It follows that

$$\Delta(2z) = \sum_{n=1}^{\infty} \tau(n)q(2z)^n = \sum_{n=1}^{\infty} \tau(n)q^{2n} = \sum_{m=1}^{\infty} \tau(\tfrac{m}{2})q^m,$$

where we define $\tau(a) = 0$ if $a$ is not an integer.

Similarly, $\Delta(z + \frac{1}{2})$ turns out to be a modular form for $\Gamma$. Because of that beautiful formula $e^{\pi i} = -1$, we have $q(z + \frac{1}{2}) = e^{2\pi i(z + \frac{1}{2})} = e^{\pi i}e^{2\pi iz} = -q(z)$, and the $q$-expansion of $\Delta(z + \frac{1}{2})$ is given by

$$\Delta\left(z + \frac{1}{2}\right) = \sum_{n=1}^{\infty}(-1)^n \tau(n)q(z)^n.$$

A more careful analysis than we have given here enables us to determine exactly which $\Gamma$ is the largest congruence group with respect to which $\theta^{24}(z)$ is a modular form of weight 12. We can then figure out the dimension of the space $M_{12}(\Gamma)$ of all modular forms of weight 12 for that $\Gamma$. Then we can figure out a basis for $M_{12}(\Gamma)$. And then we can write $\theta^{24}(z)$ as a linear combination of that basis.

It turns out that $\theta^{24}(z)$ is a linear combination of $\Delta(z + \frac{1}{2})$, $\Delta(2z)$, and an Eisenstein series of weight 12 called $E_{12}^*$. This is not the Eisenstein series $E_{12}$ we saw in chapter 13. It is similar but more complicated, because $\Gamma$ has more than one cusp, so there is more scope for different Eisenstein series. However, they all have the same flavor. Remember that $E_k$ had a $q$-expansion where the coefficient of $q^n$ was built out of the $(k-1)$st powers of the divisors of $n$. This is the flavor we are talking about. The $q$-expansion of $E_{12}^*$ looks like this:

$$E_{12}^*(z) = 1 + c \left( \sum_{n=1}^{\infty} \sigma_{11}^*(n) q^n \right).$$

The coefficients here are defined by

$$\sigma_{11}^*(n) = \begin{cases} \sigma_{11}^e(n) - \sigma_{11}^o(n) & \text{if } n \text{ is even} \\ \sigma_{11}(n) & \text{if } n \text{ is odd,} \end{cases}$$

where $\sigma_{11}^o(n)$ is the sum of the eleventh powers of the odd positive divisors of $n$, $\sigma_{11}^e(n)$ is the sum of the eleventh powers of the even positive divisors of $n$, $\sigma_{11}(n)$ is the sum of the eleventh powers of all of the positive divisors of $n$, and $c$ is a certain rational number.

So there are rational numbers $s$, $t$, and $u$ such that

$$\theta^{24}(z) = s\Delta\left(z + \frac{1}{2}\right) + t\Delta(2z) + uE_{12}^*(z).$$

You can figure out what $s$, $t$, and $u$ have to be by matching up the first few coefficients of the $q$-expansions on both sides. After doing this, you can write down the equalities that are implied by this formula for the coefficient of $q^n$ in the $q$-expansions on both sides. Because the coefficient of $q^n$ in $\theta^{24}$ is exactly $r_{24}(n)$, the number of ways of writing $n$ as a sum of 24 squares, you get the classic formula for this number, first proved by Ramanujan:

$$r_{24}(n) = \frac{16}{691}\sigma_{11}^*(n) - \frac{128}{691}\left(512\tau\left(\frac{n}{2}\right) + (-1)^n 259\tau(n)\right). \qquad (15.1)$$

The first amazing thing to notice is that because the left-hand side is an integer, the fractions on the right-hand side must add up to an integer for any $n$. You can also see the mysterious number 691 popping up. It is the denominator of the Bernoulli number $B_{12}$. It has a long history in the number theory of the last 150 years.

You might want an approximate formula for $r_{24}(n)$. We view the divisor function $\sigma^*$ as "known"—it is easy to understand and easy to compute. It turns out to be the main term, in the sense that as $n \to \infty$,

$$r_{24}(n) \sim \frac{16}{691} \sigma_{11}^*(n).$$

Asking just how good is this approximation is the same as asking how slowly does $\tau(n)$ grow with $n$ compared to $\sigma_{11}^*(n)$. There is an "easy bound" that follows from one of Jacobi's formulas right away. Ramanujan, Hardy, Rankin, and other people kept getting better bounds. The best possible bound, conjectured by Ramanujan, was proven by Deligne. It says that if any $\epsilon > 0$ is given, then there is a constant $k$ (depending on $\epsilon$) such that $\tau(n) < k n^{\frac{11}{2} + \epsilon}$. Because it is not too hard to show that there is a positive constant $k'$ such that $\sigma_{11}^*(n) > k' n^{11}$, you can see how good this approximation is.

If you compare our formula (15.1) for $r_{24}(n)$ with the formula we gave earlier for $r_2(n)$, you can see a difference. The latter was given purely in terms of functions that had a simple number-theoretic meaning, namely the number of divisors of $n$ congruent to 1 and 3 modulo 4. There are similar formulas for $r_4(n)$, $r_6(n)$, and so on for a while, involving sums of powers of various divisors $n$ and some other functions with elementary number-theoretic meaning. But when you get to $r_{24}(n)$, you get this new kind of function, $\tau(n)$.

"The function $\tau(n)$ has been defined only as a coefficient and it is natural to ask whether there is any reasonably simple 'arithmetical' definition, but none has yet been found." Hardy wrote this in 1940. In the early 1970s, Deligne proved that there was a Galois representation[5] for which the characteristic polynomials of Frobenius yield the values $\tau(p)$ for prime $p$. (It was already known how to derive the other values of $\tau(n)$ from these.) This is the "arithmetic" definition Hardy was looking for. Of course, Hardy may not have thought this was "reasonably simple."

## 3. Numerical Example and Philosophical Reflection

Let's test (15.1) and see if it works for $r_{24}(6)$. This will entail a fair amount of arithmetic, but it reveals how the formula works.

---

[5] See chapter 17 for more on Galois representations and characteristic polynomials of Frobenius.

First, on the left-hand side, let's work out $r_{24}(6)$. There are only two "basic" ways of writing 6 as a sum of squares: $6 = 1 + 1 + 1 + 1 + 1 + 1$ (shades of *Through the Looking-Glass*) and $6 = 1 + 1 + 4$. But this is not the official way to count. We have to keep track of order and signs. Imagine we have 24 numbered slots. Into each slot, we place 0, 1, $-1$, 2, or $-2$ in such a way that the sum of the squares of all the numbers in all the slots is 6. Of course, most of the slots will have to be filled with 0.

First, let's deal with the $6 = 1 + 1 + 4$ possibility. There are 24 choices for the slot in which to place 2 or $-2$. This gives 48 total choices.

For each of these choices, we have to place a 1 or $-1$ in two of the remaining 23 slots, and 0's in the rest of the slots. If we use two 1's, the number of ways of doing this is the binomial coefficient[6] $\binom{23}{2}$, read aloud as "23 choose 2." The formula gives $\binom{23}{2} = \frac{23 \cdot 22}{2}$.

The same story holds for two $-1$'s: There will be $\binom{23}{2}$ ways of putting them into the slots. If we use one 1 and one $-1$, we have 23 slots for the 1, and after we choose that slot, we have 22 left for the $-1$, giving a total of $23 \cdot 22$ possibilities. Then the rest of the slots have to be filled with 0's. In sum, the number of ways of writing 6 as a sum of squares, one of which is 4, and counted in our official manner is

$$A = 48 \left( 2\binom{23}{2} + 23 \cdot 22 \right) = 48576.$$

Next we consider the case of six $\pm 1$'s. If all the numbers have the same sign, we get $\binom{24}{6}$ ways of placing them into the slots. If we use one 1 and the rest $-1$'s, we have $\binom{24}{1}$ slots for the 1 and $\binom{23}{5}$ ways of distributing the $-1$'s into the remaining slots, and of course the

---

[6] Review of binomial coefficients: $\binom{a}{b}$ is the number of ways of putting $b$ identical things into $a$ slots. There are $a$ choices for the first slot, $a - 1$ choices for the next, down to $a - b + 1$ choices for the last. This gives us $a(a - 1) \cdots (a - b + 1)$ total choices for putting the $b$ things in the slots. But if we cannot distinguish the $b$ things because they are all identical, then the order in which we placed them in the slots is irrelevant, so we have to divide by $b!$, which is the number of all possible permutations of $b$ things. Thus we have the formula
$$\binom{a}{b} = \frac{a(a - 1) \cdots (a - b + 1)}{b!} = \frac{a!}{b!(a - b)!}.$$

computation is the same if we use one $-1$ and the rest 1's. We leave it to you to figure out what happens if there are two or three 1's and the rest $-1$'s. Notice there is a little tricky point when there are exactly three of each. In that case, the factor of 2 is missing. (Why?)

In this way, we determine that the number of ways of writing 6 as a sum of squares, all of which are $\pm 1$, counted in our official manner, is

$$B = 2\binom{24}{6} + 2 \cdot \binom{24}{1}\binom{23}{5} + 2 \cdot \binom{24}{2}\binom{22}{4} + \binom{24}{3}\binom{21}{3}.$$

Looking up the binomial coefficients or computing them, we find $B = 8614144$. The grand total is

$$r_{24}(6) = A + B = 8662720.$$

Now let's look at the right-hand side of (15.1) for $n = 6$:

$$\frac{16}{691}\sigma_{11}^*(6) - \frac{128}{691}(512\tau(3) + (-1)^6 259\tau(6)).$$

Before we do the detailed calculation, let's see if we are on the right track and haven't made any egregious errors in calculation. The biggest term is going to be

$$\frac{16}{691}6^{11} = \frac{5804752896}{691} = 8400510.70333\ldots.$$

The answer here is between 8 million and 9 million, and so was our value for $r_{24}(6)$. So far, so good.

Now let's work out all the terms.

$$\sigma_{11}^*(6) = \sigma_{11}^e(6) - \sigma_{11}^o(6) = 2^{11} + 6^{11} - 1^{11} - 3^{11} = 362621956.$$

So

$$\frac{16}{691}\sigma_{11}^*(6) = \frac{5801951296}{691} = 8396456.28944\ldots.$$

You can see just how close this "dominant" term is to the true answer of $r_{24}(6) = 8662720$. Now we compute the "error" terms.

We look up a table of the $\tau$-function and find $\tau(3) = 252$ and $\tau(6) = -6048$. So

$$-\frac{128}{691}(512\tau(3) + 259\tau(6)) = -\frac{128}{691}(512 \cdot 252 - 259 \cdot 6048)$$

$$= -\frac{128}{691}(-1437408) = 266263.710564\ldots.$$

When we add this to the dominant term, we get the *integer* (as necessary, but still surprising)

$$8396456.28944\ldots + 266263.710564\ldots = 8662720,$$

which is exactly correct.

Now for the philosophical comment. On the one hand, the way all this arithmetic works out correctly is amazing. On the other hand, it has to work out—someone proved it. (In fact, if truth be told, on our first attempt we didn't get the same number for the two sides of the equation for $r_{24}(6)$. Looking at the difference, we saw it was equal to $48 \cdot 23 \cdot 22$. This told us we must have erred when computing $A$—in fact we had made a stupid mistake—whereas we carried out the more complicated calculations correctly the first time!)

So in some sense the proved formula is "more true" than any particular calculation meant to check it. But that is so only because the formula is correct, which we know because the proof was correct. Some of the most excruciating moments in a mathematician's life occur when she thinks she has proved something but it doesn't work out in an example. Which is wrong, the proof or the example?[7]

Now, if you asked someone, "In how many ways can you write 6 as a sum of squares?" he would probably say, "Two ways: $1 + 1 + 1 + 1 + 1 + 1$ and $1 + 1 + 4$." Cavemen pondering the problem of sums of squares would no doubt have been thinking along these lines. This is a perfectly good way to start—after all, when counting partitions we don't care about the order of the parts and we don't allow 0's. But it turned out that putting the question this way

---

[7] Notice that it does not make sense to take a "postmodern" view that they can both be right, from different points of view.

does not lead to a nice theory or a nice answer in general. Instead, we rephrase the question, specifying the number of squares and counting different orders as different, and keeping track of signs of the square roots and allowing 0's. Then we get a very beautiful theory indeed, involving elliptic functions and modular forms.

After a while, something else happens. Many mathematicians get very interested in modular forms. The direct interest in sums of squares wanes. It becomes more of a historical interest or a motivating problem for studying modular forms. As the theory of modular forms advances, its new power can be tested on these old problems. In particular, various properties of partitions have been brilliantly discovered in recent years using modular forms and mock modular forms. A guru in this area is Ken Ono; see Ono (2015).

Still, we think it is fair to say that the center of interest has shifted to the modular forms themselves. Many other properties and applications of modular forms have been discovered. We mention a few of them in the next two chapters. Some of these applications are rather abstruse, for instance applications to Galois representations. Some are surprisingly concrete, for instance applications to the congruent number problem. There is an ebb and flow in interest between the motivating problems and the theory, which can get to be very complicated and abstract. The motivating problems, especially those famous in the history of number theory, become benchmarks of how powerful the theory has become. The prime example of this in our lifetimes is the proof of Fermat's Last Theorem by Wiles and by Taylor and Wiles, which used modular forms in an absolutely essential way.[8] And no other proof of Fermat's Last Theorem has yet been found.

---

[8] You can see our account of the congruent number problem in Ash and Gross (2012), and the proof of Fermat's Last Theorem in Ash and Gross (2006).

# Chapter 16

# MORE THEORY OF MODULAR FORMS

## 1. Hecke Operators

Not all modular forms are created equal. From this point on in our book, to get further into the world of modular forms, we need to single out those called *newforms*. In some books, these are called "primitive forms" or "normalized newforms."

The definition is the following. We give it in terms that we haven't discussed yet, so this definition will serve as a guide throughout this section and the next.

**DEFINITION**: A *newform* is

(a) a modular cusp form $f$ for $\Gamma_1(N)$ and level $k$ (for some $N$ and $k$) such that

(b) $f$ is normalized,

(c) $f$ is an eigenform for all the Hecke operators, and

(d) $f$ is not in the space of old forms of level $N$.

In this section, we will review the meaning of (a) and explain (b) and (c). In the next section, we will explain (d).

First, let's remember that $\Gamma_1(N)$ is the subgroup of $\mathrm{SL}_2(\mathbf{Z})$ consisting of all the matrices

$$\gamma = \begin{bmatrix} a & b \\ c & d \end{bmatrix},$$

where $a$, $b$, $c$, and $d$ are all integers, the determinant $ad - bc = 1$, $c \equiv 0 \pmod{N}$, and $a \equiv d \equiv 1 \pmod{N}$.

It is one of the mysteries, or facts, of the theory that these particular easy-to-define congruence subgroups are the main ones that are needed to do most of the work that number theorists demand from modular forms.

In particular, we know that $f$ transforms under $\gamma$'s this way:

$$f(\gamma(z)) = (cz + d)^k f(z), \qquad \text{where} \qquad \gamma(z) = \frac{az + b}{cz + d}$$

for all $z$ in the upper half-plane $H$ and all $\gamma$ in $\Gamma_1(N)$. When $f$ satisfies this transformation law, we say $f$ has "level $N$ and weight $k$."

If $f$ is a modular form of level $N$, it has a $q$-expansion

$$f(z) = a_0 + a_1 q + a_2 q^2 + \cdots + a_n q^n + \cdots,$$

where $q = e^{2\pi i z}$, and the coefficients $a_i$ are complex numbers (depending on $f$ of course). This is because the matrix $\left[\begin{smallmatrix} 1 & 1 \\ 0 & 1 \end{smallmatrix}\right]$ is in $\Gamma_1(N)$, which implies that $f(z + 1) = f(z)$. Because $f$ is periodic of period 1, it has a $q$-expansion, where $q$ relates to the cusp at $i\infty$.

There is a similar kind of expansion at each cusp. The cusps, which depend only on $N$, are points where a fundamental domain for $\Gamma_1(N)$ in $H$ "narrows down" to a point in $\mathbf{R}$ or to $i\infty$. We say that $f$ is a *cusp form* if $a_0 = 0$ and in addition there is a similar vanishing of $f$ at each of the other cusps.

Now for (b), which is easy. If $f$ is a cusp form, then its $q$-expansion has no constant term. It looks like

$$f(z) = a_1 q + a_2 q^2 + \cdots + a_n q^n + \cdots.$$

The basic number-theoretic properties of $f$ shouldn't depend too much on whether we use $f$ or $23f$ or any other nonzero multiple of $f$. It turns out that the properties of newforms are most shiningly displayed if we make sure that $a_1 = 1$. Now we can easily attain this *if* $a_1 \neq 0$, for then we can divide by $a_1$. If $a_1 = 0$, we might be perplexed. Leaving this point in the back of your mind, simply *define f* to be *normalized* if $a_1 = 1$, so the $q$-expansion of a normalized cusp form looks like

$$f(z) = q + a_2 q^2 + \cdots + a_n q^n + \cdots.$$

Now for (c). First of all, the word "operator" is just a synonym for "function." "Function" is our most general term for a rule that sends each element of one set (the source) to an element of another set (the target). When the source and target are the same set, often we use the term "operator" instead of "function." This happens particularly if the source and target is itself some set of functions, for then it becomes awkward to speak of "a function of functions." We saw another example of this elegance of diction in the term "modular form." Again, "form" is a synonym for "function," whose use is restricted to certain contexts by tradition.

So a Hecke operator $T$ is a certain function from some space of functions to itself. What space do you think will be the source and target? We will use the space of all cusp forms of level $N$ and weight $k$, which we will call $S_k(\Gamma_1(N))$. Similar to $S_k$, $S_k(\Gamma_1(N))$ is a finite-dimensional complex vector space. That means that we can choose a finite list of modular forms $f_1, \ldots, f_t$ in $S_k(\Gamma_1(N))$ with the property that *any* modular form $g$ in $S_k(\Gamma_1(N))$ can be written in a *unique* way as a linear combination of them,

$$g = b_1 f_1 + \cdots + b_t f_t,$$

for some complex numbers $b_1, \ldots, b_t$, which depend on $g$ of course and are uniquely determined by $g$ and the choice of the "basis" $f_1, \ldots, f_t$.

So far, we are looking for a Hecke operator that will be a function $T$:

$$T : S_k(\Gamma_1(N)) \to S_k(\Gamma_1(N)).$$

By the way, these operators were first discovered or put into play by Mordell, but it was Hecke who ran with the ball, and they are now named after him.

We will give a definition that is rather cut and dried and not very informative, but one that is very easy to state and compute with. The deeper reasons for this definition will have to be omitted from our book because they would require too many new ideas.

In fact, for simplicity, we are only going to give the definition of Hecke operators in the case of level $N = 1$. The definition for general level $N$ is of the same flavor, just a bit more complicated.

So let $f \in S_k(\Gamma_1(1))$ be a modular cusp form of weight $k$ and level 1. For example, if $k = 12$, $f$ could be $\Delta$. There is one Hecke operator $T_n$ for each positive integer $n$. To see what it does to $f$, take the $q$-expansion of $f$:

$$f(z) = a_1 q + a_2 q^2 + \cdots + a_s q^s + \cdots.$$

Define $T_n(f)$ by giving its $q$-expansion

$$T_n(f)(z) = b_1 q + b_2 q^2 + \cdots + b_s q^s + \cdots$$

by the formula

$$b_m = \sum_{\substack{r|n \\ r|m}} r^{k-1} a_{nm/r^2},$$

where the sum runs over all positive integers $r$ that divide both $n$ and $m$. It's not obvious, but it's true that $T_n(f)$ is again a modular form of weight $k$ and level 1.

Let's first see what happens when $m = 1$. In that case, the condition that $r|m$ forces $r$ to be 1. The sum has only one term, and we see that $b_1 = a_n$.

Another case that is not too complicated is if $n$ is a prime $p$. Then the only possible values for $r$ are 1 and $p$. We get

$$b_m = \begin{cases} a_{pm} + p^{k-1} a_{m/p} & \text{if } m \text{ is divisible by } p \\ a_{pm} & \text{otherwise.} \end{cases}$$

We can package this back into the $q$-expansion to write

$$T_p(f)(z) = \sum_{m \geq 1} a_{pm} q^m + p^{k-1} \sum_{m \geq 1} a_m q^{pm}.$$

You can see that the formula for $T_p$ plays around with the exponents and coefficients of the $q$-expansion in a precise way, involving the prime $p$ and whether $p$ divides the integer $m$.

If you like, you can just take the existence of the Hecke operators $T_n$ for granted. For the rest of this book, we will not need the exact formula that defines them. But we will need a number of facts about them.

FIRST FACT: If someone tells you exactly what $T_p$ does to a modular form $f$ in $S_k(\Gamma_1(N))$ for every prime $p$, then you can figure out what $T_n$ does to $f$, for any $n$. There are formulas that do this that are a little complicated but explicit enough to program on a computer or even to do by hand if $n$ is not too big.

SECOND FACT: Assuming $S_k(\Gamma_1(N)) \neq 0$, there are simultaneous eigenforms for all the Hecke operators, which are called *Hecke eigenforms*. What does this mean? A cusp form $f$ in $S_k(\Gamma_1(N))$ is a Hecke eigenform if $f$ is not the zero function and the "line" through $f$ doesn't budge under any of the $T_n$'s. This means that for every $n$ there is a complex number $\lambda_n$ with the property

$$T_n(f) = \lambda_n f.$$

The complex number $\lambda_n$ is called an *eigenvalue*. The "line" through $f$ is by definition the set of all complex multiples of $f$.

We can work out the implications a bit further. The equation $T_n(f) = \lambda_n f$ means that $b_m = \lambda_n a_m$ for all positive integers $m$. In particular, $b_1 = \lambda_n a_1$. However, we worked out earlier that $b_1 = a_n$. We conclude that $a_n = \lambda_n a_1$ for every positive integer $n$.

The term "eigenform" is taken from linear algebra, where any *nonzero* vector that transforms to a multiple of itself under a linear transformation is called an *eigenvector* for that linear transformation. If you start with a complex vector space $V$ of dimension greater than 1 and a linear transformation $T : V \to V$, then you can see it is usually a very special condition to be an eigenvector. Nevertheless, it is always true that a linear transformation $T$ has eigenvectors (provided that the dimension of $V$ is not 0).

So the fact that $T_n$ has an eigenform is not surprising once you know basic linear algebra. The fact that there are simultaneous eigenvectors for *all* the $T_n$'s is more surprising, but follows fairly easily from the fact that $T_n \circ T_m = T_m \circ T_n$ for any $n$ and $m$. Here the symbol $\circ$ stands for composition of functions, so we are asserting that $T_n(T_m(f)) = T_m(T_n(f))$ for any $n$ and $m$ and any cusp form $f$.

Going back to a general vector space $V$ and a linear transformation $T : V \to V$, we say that a bunch of vectors of $V$ constitute an *eigenbasis* of $V$ with respect to $T$ if each of them is an eigenvector and together they form a basis[1] of $V$. For $V$ to have an eigenbasis of

---

[1] Remember that a basis of $V$ is a set $S$ of vectors in $V$ such that any vector of $V$ is a linear combination of elements of $S$ in a unique way.

$V$ with respect to $T$ is a nontrivial condition on $T$. It is not always possible to find an eigenbasis, but when it is possible we are very happy because $T$ looks so simple when viewed through its action on an eigenbasis. The matrix of $T$ with respect to an eigenbasis is diagonal, and that's as simple a matrix as you can ask for in this context.

Unfortunately, in general it is not true that $S_k(\Gamma_1(N))$ possesses an eigenbasis for every $T_n$. We will get back to this when we discuss (d).

THIRD FACT: Suppose $f$ is a simultaneous eigencuspform for all the Hecke operators:

$$T_n(f) = \lambda_n f.$$

Suppose $f$ has the $q$-expansion

$$f(z) = a_1 q + a_2 q^2 + \cdots + a_n q^n + \cdots.$$

Then we have a very nice interplay between the Hecke eigenvalues $\lambda_n$ and the coefficients $a_n$ of the $q$-expansion, namely

$$a_1 \lambda_n = a_n$$

for all $n$, which we derived above.

It follows from this equation that $a_1 \neq 0$ because by definition an eigenform is a nonzero modular form. So in this case we can always normalize such an $f$. Let's say that $f$ is a *normalized Hecke eigenform* if it is a cusp form that is a simultaneous eigenform for all the Hecke operators and whose leading term in its $q$-expansion is $q$.

We can summarize by saying that if $f$ is a normalized Hecke eigenform, then

$$f(z) = q + a_2 q^2 + \cdots + a_n q^n + \cdots$$

and

$$T_n(f) = a_n f$$

for all $n$. This is very nifty.

Because a modular form is determined by its $q$-expansion, this implies that a normalized Hecke eigenform is determined by its Hecke eigenvalues. If you have $f$ and $g$ with the same Hecke eigenvalues right down the line, then $f = cg$ for some constant $c$. If both of them are normalized, then of course $c = 1$ and $f = g$. Because all the Hecke operators can be computed in terms of the $T_p$'s for prime $p$, we see that if two normalized Hecke eigenforms have the same $a_p$'s in their $q$-expansions for all primes $p$, then they must be equal, meaning that they have the same coefficients $a_n$ for all $n$.

FOURTH FACT: Suppose $f$ is a normalized Hecke eigenform with Hecke eigenvalues $a_n$. Therefore the $q$-expansion of $f$ is

$$f(z) = q + a_2 q^2 + \cdots + a_n q^n + \cdots.$$

Because the Hecke operator $T_n$ can be computed in terms of the $T_p$'s for $p$ prime, it follows that there must be a formula for $a_n$ in terms of the $a_p$'s. We won't write down the general formula for this, but you can deduce it from things we say in the third section of this chapter. It is not surprising perhaps that $a_n$ depends on the $a_p$'s only for those primes $p$ that divide $n$. It turns out in particular that if $n$ and $m$ are relatively prime, meaning that they share no prime factors, then

$$a_{nm} = a_n a_m.$$

This is a very nice property, expressed by saying that the function $n \to a_n$ is *multiplicative*. Notice that in this context multiplicativity does not mean $a_{nm} = a_n a_m$ for every two numbers $n$ and $m$, only those integers $n$ and $m$ that are relatively prime.

For example, $\Delta$ is a normalized Hecke eigenform[2] of weight 12 and level 1. Its Hecke eigenvalues are $a_n = \tau(n)$, the Ramanujan $\tau$-function. See the values of $\tau(n)$ given in section 4 of chapter 13, where we remarked that $\tau(6) = \tau(2)\tau(3)$.

Thus the $\tau$-function is multiplicative. If we fed this back into the formula for the number $r_{24}(n)$ of representations of an integer $n$ as a sum of 24 squares, we could write down some strange-looking and *a priori* totally unpredictable formulas relating the values of $r_{24}(n)$ for various $n$.

---

[2] It has to be an eigenform because the dimension of the $S_{12}(1)$ happens to be exactly 1. There is nowhere for $\Delta$ to go when applying Hecke operators except to multiples of itself.

## 2. New Clothes, Old Clothes

Now it is time to explain (d). We'd like to work with a subspace of $S_k(\Gamma_1(N))$ that *does* have a simultaneous Hecke eigenbasis. It turns out we can do that without really losing anything. Here's how.

There are ways to take modular forms of level $M$ and alter them so that they become level $N$ if $N$ is a multiple of $M$. The easiest way to do this is just to stare at the modular form twice as hard as you were staring. In other words, suppose $f(z)$ has weight $k$ and level $M$. Assume that $N$ is a multiple of $M$, strictly larger than $M$. Thus $f(\gamma(z)) = (cz + d)^k f(z)$ for all $\gamma$ in $\Gamma_1(M)$. But a little exercise shows that $\Gamma_1(N)$ is a subgroup of $\Gamma_1(M)$. So this same transformation formula holds in an utterly tautological fashion for every $\gamma$ in $\Gamma_1(N)$. The other conditions to be a modular form also continue to hold for $f(z)$. Thus $f(z)$ can also be considered to have level $N$.

The modular form $f(z)$ shouldn't "really" have level $N$, but it does according to our definitions. It is a member both of $S_k(\Gamma_1(M))$ and of $S_k(\Gamma_1(N))$. So we call it an "old form" when we are looking at it in $S_k(\Gamma_1(N))$.

We can be a bit more clever. If $N > M$ is a multiple of $M$ and if $t$ is any divisor of $N/M$, and if $f(z)$ has level $M$, then it is not too hard to check that $f(tz)$ has level $N$. We call all modular forms obtained this way "old forms" of level $N$. The set of all their linear combinations is called the "old subspace" of $S_k(\Gamma_1(N))$. Call this $S_k(\Gamma_1(N))^{\text{old}}$. We also call any element of $S_k(\Gamma_1(N))^{\text{old}}$ an old form. Any number theory coming from old modular forms can already be studied at the lower levels they "come from."

FIFTH FACT: Any modular form of level $N$ can be written as a linear combination of some newforms and an old form. The newforms are each simultaneous Hecke eigenforms with multiplicative coefficients for their $q$-expansions. The old form is considered "understood" already by investigating its components in their natural habitat at smaller levels.

We already mentioned the standard example: $\Delta$ is a newform. It can't be old, because it has level 1, so there is no smaller level for it to come from.

For other examples, look at weight 2. We know that $S_2(\Gamma_1(1)) = 0$.

There are no modular forms of weight 2 and level 1. Now, suppose $N$ is prime, so that its only strictly smaller divisor is 1. Then again there are no old forms in $S_2(\Gamma_1(N))$, because the only smaller level they could come from is level 1, and there is nothing there. So if $N$ is prime, $S_2(\Gamma_1(N))$ has a basis of newforms.

For example, if $N = 11$, it happens that $S_2(\Gamma_1(11))$ is one-dimensional, so any nonzero member of it can be normalized to give the unique normalized newform of weight 2 and level 11. Here is the $q$-expansion of this newform:

$$f(z) = q - 2q^2 - q^3 + 2q^4 + q^5 + 2q^6 - 2q^7 - 2q^9$$
$$- 2q^{10} + q^{11} - 2q^{12} + 4q^{13} + \cdots .$$

Although we will mostly be talking about newforms from here on out, old forms have various important uses in number theory as well. One of them we have already seen. The representation numbers $r_{24}(n)$ are the coefficients of the $q$-expansion of an old form $\Phi$ of level 4 plus an Eisenstein series. If you look back at the previous chapter, you will see that this form $\Phi$ is a linear combination of $\Delta(2z)$ and $\Delta(z + \frac{1}{2})$. The first of these is obviously an old form of level 4 but so is the second.[3]

---

[3] The $q$-expansion of $\Delta$ is $\sum \tau(n)q^n$. If you look back in the previous chapter at the $q$-expansion of $\Delta(z + \frac{1}{2})$, you will see that

$$\frac{1}{2}\left(\Delta(z + \frac{1}{2}) + \Delta(z)\right) = \sum \tau(2n)q^{2n}$$

so

$$\Delta\left(z + \frac{1}{2}\right) = -\Delta(z) + 2\sum \tau(2n)q^{2n}.$$

But $\Delta$ is a Hecke eigenform. In particular, $T_2(\Delta) = -24\Delta$. Written out in terms of $q$-expansions, this becomes

$$\sum \tau(2n)q^n + 2^{11} \sum \tau(n)q^{2n} = -24\Delta.$$

Now replace $z$ by $2z$. We get

$$\sum \tau(2n)q^{2n} + 2^{11} \sum \tau(n)q^{4n} = -24\Delta(2z),$$

or in other words

$$\sum \tau(2n)q^{2n} = -2^{11}\Delta(4z) - 24\Delta(2z).$$

Substituting in the second equation above, we get

$$\Delta\left(z + \frac{1}{2}\right) = -\Delta(z) - 48\Delta(2z) - 2^{12}\Delta(4z).$$

We'd like to thank David Rohrlich for this clever derivation.

## 3. *L*-functions

Mathematicians have defined $\zeta$-functions and *L*-functions "attached" to many different mathematical objects. They are Dirichlet series whose coefficients are computable from the properties of the given mathematical object. Whether we call the resulting function by the letter $\zeta$ or by *L* is a matter of tradition.

When two different objects have the same *L*-function, this can mean there is a very profound and often very useful tight connection between the two objects. We will see examples of this in the next chapter. In this section, we want to explain how you "attach" an *L*-function to a modular form, and we list some of the extraordinary properties of these *L*-functions.

Suppose you have an infinite list of complex numbers $a_1, a_2, \ldots$. What can you do with them? You could make them into a power series

$$a_1 q + a_2 q^2 + \cdots$$

and ask whether this is the $q$-expansion of a modular form. Or you could make them into a Dirichlet series

$$\frac{a_1}{1^s} + \frac{a_2}{2^s} + \cdots$$

and ask whether this gives rise to an analytic function of $s$ in some domain—hopefully the whole complex plane. In particular, if the coefficients $a_n$ don't grow too fast with $n$, then the series defines an analytic function on some right half-plane, and then we can ask whether that function can be analytically continued further to the left.

Now we can put these two ideas together. If we start with a modular newform $f(z)$ with $q$-expansion

$$f(z) = q + a_2 q^2 + \cdots,$$

then we can use those $a_n$'s to form a Dirichlet series. Since this series depends on what $f$ we start with, we put it into the notation. We use the notation

$$L^*(f, s) = 1 + \frac{a_2}{2^s} + \cdots.$$

The reason for the asterisk is because this is not the official $L$-function of $f$. For the sake of neatness, we will modify $L^*$ a bit to get the official $L$.

Conversely, if we start with some Dirichlet series (which is often an $L$-function of some other object such as an elliptic curve or a Galois representation), then we can form the $q$-expansion with the same $a_n$'s and ask if it is the $q$-expansion of a modular form.

Suppose now that $f$ is a newform of weight $k$ and level $N$. Define

$$L(f,s) = N^{s/2}(2\pi)^{-s}\Gamma(s)L^*(f,s).$$

Here, $\Gamma(s)$ is the $\Gamma$-function discussed in chapter 7. Because $N$ and $2\pi$ are positive numbers, there is no problem about raising them to a complex power.

Hecke proved some amazing facts about $L(f,s)$. First, although it starts life only defined on some right half-plane $\mathrm{Re}(s) > t_0$ (where the Dirichlet series converges absolutely and the $\Gamma$-function is analytic), it can be continued[4] to be an analytic function for all $s \in \mathbf{C}$.

Second, there is a factorization of $L^*(f,s)$ that parallels Euler's factorization of the $\zeta$-function into factors, one for each prime number $p$:

$$L^*(f,s) = \prod_{p \nmid N} \frac{1}{1 - a_p p^{-s} + \chi(p)p^{k-1}p^{-2s}} \times \prod_{p \mid N} \frac{1}{1 - a_p p^{-s}}.$$

The first product is taken over all primes $p$ that do not divide the level $N$, and the second product is taken over the $p$ that do divide $N$. The function $\chi(p)$ is a certain nice function that depends on $f$. It takes values in the roots of unity, and although it is very important in the theory (where it is called the *nebentype character of $f$*), we do not need to say more about it here.

This factorization reflects the fact that $f$ is an eigenform for all the Hecke operators. In fact, if you take the trouble to use the formula for the geometric series to do the indicated divisions, you can recover the formula that tells you what $a_{p^m}$ is in terms of $a_p$, $m$,

---

[4] We discussed analytic continuation a little bit in chapter 7 and more extensively in Ash and Gross (2012, chapter 12).

and $\chi(p)$. The factorization also tells you that the function $n \to a_n$ is multiplicative.[5]

The third fact is the "functional equation":

$$L(f,s) = i^k L(f^\dagger, k - s).$$

In this formula, $f^\dagger$ is another modular form of the same level and weight as $f$ and closely related to $f$.[6]

The existence of this functional equation shows that the coefficients $a_n$ of the $q$-expansion of $f$ are not just some random numbers but have very tight, perhaps we could say mystical, relationships among themselves. The factorization over primes $p$ also shows a relationship among the $a_n$'s, but those relationships are not so mystical. They are simply the relationships required by the fact that $f$ is an eigenform of all the Hecke operators.

---

[5] Remember that this means that if $n$ and $m$ share no prime factor, then $a_{nm} = a_n a_m$.

[6] The form $f^\dagger$ is the image of $f$ under the Atkin–Lehner operator. For more about this and about everything regarding modular forms and more, see Ribet and Stein (2011), a very nice set of notes.

# Chapter 17

# MORE THINGS TO DO WITH MODULAR FORMS: APPLICATIONS

> [W]e descended into a dismal bog called 'sums.' There
> appeared to be no limit to these. When one sum was
> done, there was always another.
> —Winston Churchill, *My Early Life*

In this last chapter, we give a small sample of other areas of number theory that benefit from the application of the theory of modular forms. The first two sections refer to Ash and Gross (2006; 2012). In each of those books, we found ourselves forced to refer to modular forms briefly. If we had been able to go more deeply into modular forms, we would have written what we are going to write here. Obviously, we cannot recapitulate the full contents of our two earlier books in this one. Therefore, we apologize for the necessary obscurity, but we wish to supplement those books with the discussion in the next two sections.

After those sections, we will mention another two applications of modular forms to interesting problems, one more a part of group theory than number theory, the other soundly ensconced in the theory of elliptic curves. We will finish with a glance at the future.

Before beginning, we should mention that of course the theory of modular forms is a very interesting subject in its own right—even without any of the applications. For example, if you have two modular forms, both of the same level $N$ but with possibly different weights $k_1$ and $k_2$, you can multiply them together to get a new modular form, again of level $N$ but now of weight $k_1 + k_2$. Thus the set of all modular forms of level $N$ and integral weight forms a "ring"—it is closed under addition and multiplication.

The structure of this ring is very interesting and is closely connected with the algebraic geometry of the Riemann surface formed as the set of orbits in the upper half-plane $H$ under the action of the congruence group $\Gamma_1(N)$. This Riemann surface is called $Y_1(N)$ and has a lot of number theory content. In fact, $Y_1(N)$ can be defined by algebraic equations in two variables with coefficients in a field of algebraic numbers.

## 1. Galois Representations

In Ash and Gross (2006), we discussed the concept of a Galois representation. To be extremely brief, a complex number is called *algebraic* if it is the root of a polynomial with integral coefficients. The set of all algebraic numbers, written $\overline{\mathbf{Q}}$, makes up a field: You can add, subtract, multiply, and divide them (as long as you don't divide by zero), and the result is still an algebraic number.

The "Absolute Galois group of $\mathbf{Q}$," called $G_{\mathbf{Q}}$, is the set of all one-to-one correspondences $\sigma : \overline{\mathbf{Q}} \to \overline{\mathbf{Q}}$ that "preserve arithmetic"—namely, $\sigma(a+b) = \sigma(a) + \sigma(b)$ and $\sigma(ab) = \sigma(a)\sigma(b)$ for all algebraic numbers $a$ and $b$ in $\overline{\mathbf{Q}}$. To say that $G_{\mathbf{Q}}$ is a group means that you can compose any two elements of it (the elements are functions from $\overline{\mathbf{Q}}$ to the same set $\overline{\mathbf{Q}}$ and so can be composed) and get a new element of $G_{\mathbf{Q}}$. Also, the inverse function of an element of $G_{\mathbf{Q}}$ is still in $G_{\mathbf{Q}}$.

The Absolute Galois group of $\mathbf{Q}$ contains a tremendous amount of information. Number theorists have been mining this information at an increasing rate over the last several generations. One way to get bits of it is to discover or consider *Galois representations*. Such a thing is a function

$$\rho : G_{\mathbf{Q}} \to \mathrm{GL}_n(K),$$

where $K$ is a field and $\mathrm{GL}_n(K)$ denotes the group of $n$-by-$n$ matrices with entries in $K$. We require that $\rho$ satisfy the rule $\rho(\sigma\tau) = \rho(\sigma)\rho(\tau)$. Here, $\sigma\tau$ represents the composition of $\tau$ first and then $\sigma$, and $\rho(\sigma)\rho(\tau)$ stands for the multiplication of the two matrices $\rho(\sigma)$ and $\rho(\tau)$. We also require $\rho$ to be continuous, a technical (but important) condition that we need not explain here.

If you have a Galois representation $\rho$, what can you do with it? If it is the same Galois representation as you got from some other venue, then the fact that the two are the same often means you have conjectured or discovered some very important relationship in number theory. Sometimes these relationships are called "reciprocity laws."

Eichler and Shimura (for weight 2), Deligne (for weight $> 2$), and Deligne and Serre (for weight 1) proved that any newform is connected to a Galois representation as follows. For simplicity, we will consider only Galois representations where the field $K$ is a finite field that contains the field with $p$ elements. (Here, $p$ can be any prime number. Because we are using $p$ for this purpose, what we used to call $p$ when talking about Hecke operators we will now call $\ell$.)

So suppose $f(z)$ is a newform of level $N$ and weight $k \geq 1$ with $q$-expansion

$$f(z) = q + a_2 q^2 + \cdots .$$

Remember that saying $f(z)$ is a newform implies that the eigenvalue of the Hecke operator $T_\ell$ is $a_\ell$. Choose a prime $p$. Then there exists a field $K$ containing the field with $p$ elements and a Galois representation

$$\rho : G_{\mathbf{Q}} \to \mathrm{GL}_2(K)$$

"attached to $f$." (Note that $n = 2$ here: We always get *two*-dimensional representations from newforms.) What does "attached" mean?

For any prime $\ell$, there are elements of $G_{\mathbf{Q}}$ called "Frobenius elements at $\ell$." There are many of them, but we use the notation $\mathrm{Frob}_\ell$ for any one of them. This may seem like a dangerous thing to do, but we will ultimately write down formulas that won't depend on which Frobenius element at $\ell$ we work with.

In general, $\rho(\mathrm{Frob}_\ell)$ does depend on which Frobenius element we use, so we will not refer to the naked matrix $\rho(\mathrm{Frob}_\ell)$. However, it turns out, wonderfully enough, that if $\ell \neq p$ and $\ell$ is not a prime factor of $N$, then the determinant $\det \rho(\mathrm{Frob}_\ell)$ does *not* depend on which Frobenius element we choose. In fact, neither does the somewhat more complicated determinant $\det(I - X\rho(\text{Frob}_\ell))$, which is a polynomial of degree 2 in $X$. It is called the *characteristic polynomial of Frobenius* under $\rho$ at $\ell$.

We say that $\rho$ is *attached* to $f$ if, for all $\ell$ not dividing $pN$, we have an equality of polynomials:

$$\det(I - X\rho(\text{Frob}_\ell)) = 1 - a_\ell X + \chi(\ell)\ell^{k-1}X^2.$$

Here, $\chi$ is the same function of primes that arose in the $L$-function of $f$ discussed in the last chapter. It's a nice function whose exact definition need not concern us.[1]

When $\rho$ is attached to $f$, we say we have a *reciprocity law* between Galois theory and the theory of modular forms. The explanation for this terminology is rather lengthy and is given in Ash and Gross (2006). Such a reciprocity law is a powerful tool that on the one hand can be used to study the absolute Galois group and Diophantine equations and on the other hand can be used to learn more about modular forms. It is a two-way street, and both directions have been very fruitful in recent number theory. Such reciprocity laws were crucial in the proof of the modularity conjecture and Fermat's Last Theorem (FLT) due to Wiles and also Taylor and Wiles. This, too, is discussed in Ash and Gross (2006).[2] We will talk about the modularity conjecture in the next section.

The theorems of Eichler and Shimura, Deligne, and Deligne and Serre just mentioned are getting old, as we are getting old along with them. They were proved between the 1950s and the 1970s. However, in more recent history, a great converse to these theorems was proved by Khare and Wintenberger (2009a; 2009b).

Here is the theorem of Khare and Wintenberger, whose proof builds on the work of Wiles, Taylor and Wiles, and many others. To state it, we have to introduce the element $c$ in the Absolute Galois group $G_\mathbf{Q}$, which is simply complex conjugation. Because the nonreal roots of an integral polynomial in one variable come

---

[1] On the right-hand side of this equation, we view $a_\ell$, $\chi(\ell)$, and $\ell^{k-1}$ as "reduced modulo a prime above $p$," so that they lie in the field $K$.

[2] A more technical treatment, with nearly all of the details that are missing from popular treatments, can be found in Cornell et al. (1997).

in conjugate pairs, $c$ takes algebraic numbers to algebraic numbers and so is an element of $G_{\mathbf{Q}}$.

**THEOREM 17.1**: *Suppose $p$ is prime and $K$ is a finite field containing $\mathbf{Z}/p\mathbf{Z}$. Given*

$$\rho : G_{\mathbf{Q}} \to \mathrm{GL}_2(K),$$

*a Galois representation with the property that $\det \rho(c) = -1$ if $p$ is odd (there is no condition if $p = 2$), then there exists a modular form $f$ with $\rho$ attached. Moreover, there is an explicit, but rather complicated, recipe for determining the level, weight, and $\chi$-function that belong to $f$.*

This theorem was conjectured by Serre in 1987. The condition on $\rho(c)$ is a necessary condition because any Galois representation attached to a newform must obey it.

The discovery and proof of this tight connection between two-dimensional Galois representations and modular forms is one of the glories of number theory in the last half-century. If you just write down the definitions of a two-dimensional Galois representation and a modular form, no particular connection between them flies off the page. Yet they control each other (at least if $\det \rho(c) = -1$). Many features of the landscape are explained starting from this connection.

## 2. Elliptic Curves

Now we go on to some of the material in Ash and Gross (2012). An elliptic curve $E$ can be given by an equation of the form

$$y^2 = x^3 + ax^2 + bx + c,$$

where $a$, $b$, and $c$ are complex numbers. If we choose them to be rational numbers, then we get an elliptic curve "over $\mathbf{Q}$." There are many interesting questions to work on here, but the most obvious is to ask about the solutions $(x, y)$ to this equation. We are especially interested in the solutions where $x$ and $y$ are both rational numbers.

But to begin to study the situation, we have to understand all the solutions where $x$ and $y$ are complex numbers. If $R$ is any system of numbers, we write $E(R)$ to stand for the set of all solutions to the equation where $x$ and $y$ lie in $R$, and we add on one extra solution called $\infty$ (which stands for $x = \infty$ and $y = \infty$—a solution that you can make sense of by using a little projective geometry).

When we do this, we find that $E(\mathbf{C})$ is a torus in shape. It turns out that this torus can be naturally viewed as a parallelogram in the complex plane with opposite sides glued together. This parallelogram has vertices at 0, 1, $z$, and $z + 1$, where $z$ is some point in the upper half-plane $H$, and the choice of $z$ gives a connection between $H$ and elliptic curves that leads straight into the theory of modular forms.

As we discussed in Ash and Gross (2012), an elliptic curve $E$ over $\mathbf{Q}$ has an $L$-function of its own, $L(E, s)$. This is an analytic function for all $s$ in $\mathbf{C}$ built out of data coming from the number of solutions modulo $\ell$, namely the size of the finite set $E(\mathbf{Z}/\ell\mathbf{Z})$, for each prime $\ell$. The main topic of Ash and Gross (2012) was the conjecture of Birch and Swinnerton-Dyer, asserting that "how many" solutions there are in $E(\mathbf{Q})$ can be related in a certain way to the properties of $L(E, s)$.

The modularity conjecture said that given any $E$ over $\mathbf{Q}$ there exists a newform $f$ of weight 2 such that $L(E, s) = L(f, s)$. (The level of $f$ is predicted from certain properties of $E$.) This conjecture was first cracked by the work of Wiles and of Taylor and Wiles referred to in the previous section, and the proof was completed in a paper by Breuil, Conrad, Diamond, and Taylor. Probably the proof of this conjecture is ultimately more significant than its by-product, the proof of FLT. However, famous problems such as FLT serve as benchmarks to measure how much we understand of number theory.

The modularity conjecture has to be assumed before you can state the conjecture of Birch and Swinnerton-Dyer. This is because the latter conjecture involves the behavior of $L(E, s)$ around the point $s = 1$. However, from the mere definition of $L(E, s)$, it is only defined as an analytic function when the real part of $s$ is greater than 2. The only known way to prove that $L(E, s)$ extends to an analytic function on the whole $s$-plane is to prove that $L(E, s) = L(f, s)$ for some newform $f$. Then, as we saw in the previous chapter, Hecke had already proved that $L(f, s)$ extends to an analytic function on the whole $s$-plane.

Once you know the modularity conjecture is true, then you can ask all kinds of interesting questions concerning the $f$ you get from $E$, and such questions give you ideas for asking new questions about $f$'s of weight other than weight 2.

## 3. Moonshine

In this section, we want to give an example that moves a bit outside of number theory. In the theory of finite groups, there is a concept of a "simple group." These groups are not necessarily so simple. The term refers to the fact that they are the building blocks of all finite groups, analogous to how the prime numbers are building blocks of all integers and how all molecules are built of elements. The finite simple groups have all been found and listed, in one of the big accomplishments of twentieth-century mathematics. There are an infinite number of them, but they can be listed in various infinite families, along with 26 groups that don't fit into any of the families. The largest of these 26 "sporadic" groups, and the last to be discovered, is named the "Monster" group, and is denoted by the letter $M$. It was proved to exist in 1982 by Robert Griess. A good book about all this is Ronan (2006).

Hold those thoughts. On the other hand, we consider $j(z)$, a weakly modular form of level 1 and weight 0 that has been known for a couple of centuries. (A *weakly modular form* is like a modular form, except that its $q$-expansion is allowed to have a finite number of negative powers of $q$.) As you can see from the definition of modular form, a modular form of level 1 and weight 0 is an analytic function on $H$ that is invariant under the modular group

$$j(\gamma(z)) = j(z)$$

for every $z$ in $H$ and every $\gamma$ in $\mathrm{SL}_2(\mathbf{Z})$. The $q$-expansion of any nonzero function with these properties must have negative exponents. In some sense, $j$ is the simplest nonzero example of all these functions, because its $q$-expansion starts with $q^{-1}$—it doesn't involve any other negative exponents. In fact,

$$j(z) = q^{-1} + 744 + 196884q + 21493760q^2 + \cdots.$$

We can construct $j$ out of modular forms of higher weight as

$$j = \frac{E_4^3}{\Delta}.$$

If $E$ is the elliptic curve $\mathbf{C}/(\mathbf{Z} + z\mathbf{Z})$, then $j(z)$ is an important number attached to $E$, called its *j-invariant*. That's why it was discovered back in the nineteenth century, as part of the study of elliptic functions.

So what do these things have to do with each other? In the late 1970s, John McKay noticed that the coefficients of the $q$-expansion of $j(z)$ were closely related to properties of the Monster group $M$. Even though $M$ had not yet been proven to exist, many of its properties (assuming it did exist) were known.

To be precise, we can explain the relevant properties like this. In general, to study any group, we can explore its "representations." These are homomorphisms[3] from the group to a group of matrices. For example, we explore the Absolute Galois group of $\mathbf{Q}$ by trying to understand Galois representations. All the representations of a group can be built from building blocks (more building blocks!) called *irreducible representations*. If we have a group, we can try to make a list of its irreducible representations and we can look at their dimensions. If

$$f : G \to \mathrm{GL}_n(\mathbf{C})$$

is a representation of a group $G$, then its dimension is $n$.

For example, every group has the trivial representation $f : G \to GL_1(\mathbf{C})$ that sends every element of $G$ to the matrix $[1]$. The trivial representation is irreducible and has dimension 1.

The irreducible representations of $M$ have the dimensions $d_1 = 1$, $d_2 = 196883$, $d_3 = 21296876, \ldots, d_m, \ldots$. What McKay noticed was

---

[3] If $G$ and $H$ are groups, a homomorphism from $G$ to $H$ is a function $f : G \to H$ with the property that $f(g_1g_2) = f(g_1)f(g_2)$ for all elements $g_1$ and $g_2$ in $G$.

a strange connection between these numbers and the coefficients of the $q$-expansion of $j$. You have to skip the constant term 744 but look at the others: The coefficient of $q^{-1}$ is $1 = d_1$, The coefficient of $q$ is $d_1 + d_2$, and the coefficient of $q^2$ is $d_1 + d_2 + d_3$. Similar formulas hold for the other coefficients in terms of the $d_m$—although the coefficients are not always equal to 1. But they tend to be small positive integers.

This strange and unexpected connection between the $j$-function and the Monster group was called "Monstrous Moonshine" by John Conway and Simon Norton. Similar connections were discovered between the coefficients of other modular forms of weight 0 and other finite groups. The study of these connections used ideas from physics and other parts of mathematics. Following the work of many mathematicians, a full explanation of Monstrous Moonshine was given by Richard Borcherds in 1992, and he won the Fields Medal partly because of this work.

## 4. Bigger Groups (Sato–Tate)

You may have noticed that we have used 2-by-2 matrices a lot in the theory of modular forms. The modular group consists of certain 2-by-2 matrices, and the Galois representations attached to modular forms are valued in 2-by-2 matrices. What about larger matrices?

For several generations now, mathematicians have been generalizing the theory of modular forms to groups of larger-sized matrices as well as to groups of other kinds of 2-by-2 matrices. Even 1-by-1 matrices can be brought in to create a uniform picture, which is called the theory of "automorphic forms." This is not the place to describe this research, but suffice it to say that this has been one of the most vibrant areas of number theory in recent years.

It turns out that even if you were only interested in 2-by-2 matrices, you could still gain a lot through doing the more general theory and applying your results to the 2-by-2 matrix theory. As an example of this, we will mention the proof of the Sato–Tate Conjecture, which used the theory of automorphic forms on larger matrix groups in an essential way. This proof was announced in 2006 by Clozel, Harris, Shepherd-Baron, and Taylor. As usual, this great work has spawned all kinds of generalizations, both conjectural and proven. The whole area of automorphic forms is now a mature and profound source of problems and tools for proof in number theory.

So what is the Sato–Tate Conjecture? Suppose $E$ is an elliptic curve over $\mathbf{Q}$. Because $E$ can be defined via an equation with integer coefficients, we can reduce the coefficients modulo any prime $\ell$ and consider the set of solutions mod $\ell$. [You have to be a little careful how you do this, and you have to include $\infty$ in your count. We discussed these issues in Ash and Gross (2012).]

Let $N_\ell$ be the number of these solutions, namely the number of elements of $E(\mathbf{Z}/\ell\mathbf{Z})$. It turns out that $N_\ell$ is fairly close to $1 + \ell$, and we call the discrepancy $a_\ell$. That is,

$$a_\ell = 1 + \ell - N_\ell.$$

Long ago, Hasse proved that the absolute value $|a_\ell|$, which of course is a nonnegative integer, is always less than $2\sqrt{\ell}$. The question is: How does $a_\ell$ vary with $\ell$? Naturally, as $\ell \to \infty$, $|a_\ell|$ can get larger and larger. To keep things within a finite playing field, we normalize by considering

$$\frac{a_\ell}{2\sqrt{\ell}},$$

which is always between $-1$ and $+1$. Now the cosine of an angle between 0 and $\pi$ radians also lies between $-1$ and $+1$, and it is traditional to define $\theta_\ell$ by the formula

$$\cos\theta_\ell = \frac{a_\ell}{2\sqrt{\ell}}, \qquad 0 \le \theta_\ell \le \pi.$$

Our question now becomes: How does $\theta_\ell$ vary with $\ell$ for a fixed elliptic curve $E$?

There are two kinds of elliptic curves, CM curves and non-CM curves,[4] and the expected answer to our question depends on which kind $E$ is. Let's assume $E$ is non-CM. Draw a semicircle in the upper

---

[4] It doesn't matter here which is which, but rest assured it is usually pretty easy to tell the kind of any particular elliptic curve $E$.

half-plane and start plotting the points on it that have angle $\theta_2$, $\theta_3$, $\theta_5$, $\theta_7$, .... (It actually is not so hard to program a computer to do this.) The points will scatter around in what appears to be a fairly random pattern, but after you plot many of them, you will see that they seem to fill out the semicircle, more thickly in the middle than on the ends. The exact distribution of these points as $\ell \to \infty$ was conjectured by Sato and by Tate (independently). Their conjecture is formulated exactly by saying that the probability distribution defined by the $\theta_\ell$'s is $\sin^2 \theta \, d\theta$.

What this means is that if you choose an arc on the semicircle, say $\phi_1 < \theta < \phi_2$, and if you compute the fraction[5] of points in that semicircle

$$f_L = \frac{\#\{\ell < L \text{ such that } \phi_1 < \theta_\ell < \phi_2\}}{\#\{\ell < L\}},$$

then the answer (for large $L$) will be very close to

$$c = \frac{2}{\pi} \int_{\phi_1}^{\phi_2} \sin^2 \theta \, d\theta,$$

and these two numbers will be equal in the limit as $L \to \infty$ (i.e., the limit as $L \to \infty$ of $f_L$ exists and equals $c$).[6]

This conjecture is now a theorem. For example, if we take a very small arc around $\pi/2$, then $\sin \theta$ is around 1, about as large as it can be, so the points cluster most thickly in this middle arc.

## 5. Envoy

The sum was done.
—*Ulysses*

We have come a long way from $2 + 2 = 4$ (we almost wrote $2 + 2 = 5$). Mathematics and number theory in particular are the

---

[5] Of course, if we plot the points for *all* the primes $\ell$, we will get infinitely many points in any arc. Instead we limit ourselves to all $\ell$ less than some large number $L$ and then let $L$ tend to infinity.

[6] Check this makes sense by computing the integral $\int_0^\pi \sin^2 \theta \, d\theta$ and getting $\pi/2$. HINT: $\sin^2 \theta + \cos^2 \theta = 1$.

paradigms of solid human knowledge, even if modern philosophers can pick holes in this solidity. No one has ever found a contradiction in number theory. (Of course, when someone finds what appears to be a contradiction, she works hard until she finds some mistake in the reasoning. She does not persist after that to find new mistakes.)

The field of questions number theorists find interesting develops with the theory. Problems that are very difficult for well-understood structural reasons, for instance writing down nice formulas for the number of ways of writing an integer as a sum of a large *odd* number of squares, tend to move to the back of the line. Problems whose solutions expose new structures are favored, and often those structures become central objects of curiosity and research themselves. For example, the Sato–Tate Conjecture goes into rather fine detail concerning the number of solutions of a cubic equation modulo primes. Sato and Tate wouldn't even have made the conjecture if someone hadn't first discovered the bound $|a_\ell| < 2\sqrt{\ell}$.

Meanwhile, the theory of automorphic representations, used to prove the Sato–Tate Conjecture, had already become a major field in number theory by itself. The fact that it could be used to prove the conjecture is a sign of how powerful that theory is. The questions provoke the theory, the theory provides new questions, and the conjectures provide guideposts along the way. We keep on asking more and more difficult questions, whose solutions can measure our progress in a sort of war against ignorance. As André Weil described it (Weil, 1962, Introduction): "this bloodless battle with an ever retreating foe which it is our good luck to be waging."

# Bibliography

Ash, Avner, and Robert Gross. *Elliptic Tales: Curves, Counting, and Number Theory*, Princeton University Press, Princeton, NJ, 2012.

———. *Fearless Symmetry: Exposing the Hidden Patterns of Numbers*, Princeton University Press, Princeton, NJ, 2006. With a foreword by Barry Mazur.

Bell, E. T. *Men of Mathematics*, Simon and Schuster, New York, 1965.

Boklan, Kent D., and Noam Elkies. *Every Multiple of 4 Except 212, 364, 420, and 428 is the Sum of Seven Cubes*, February, 2009, http://arxiv.org/pdf/0903.4503v1.pdf.

Buzzard, Kevin. *Notes on Modular Forms of Half-Integral Weight*, 2013, http://www2.imperial.ac.uk/~buzzard/maths/research/notes/modular_forms_of_half_integral_weight.pdf.

Calinger, Ronald. *Classics of Mathematics*, Pearson Education, Inc., New York, NY, 1995. Reprint of the 1982 edition.

Cornell, Gary, Joseph H. Silverman, and Glenn Stevens. *Modular Forms and Fermat's Last Theorem*, Springer-Verlag, New York, 1997. Papers from the Instructional Conference on Number Theory and Arithmetic Geometry held at Boston University, Boston, MA, August 9–18, 1995.

Davenport, H. *The Higher Arithmetic: An Introduction to the Theory of Numbers*, 8th ed., Cambridge University Press, Cambridge, 2008. With editing and additional material by James H. Davenport.

Downey, Lawrence, Boon W. Ong, and James A. Sellers. "Beyond the Basel Problem: Sums of Reciprocals of Figurate Numbers," *Coll. Math. J.*, 2008, **39**, no. 5, 391–394, available at http://www.personal.psu.edu/jxs23/downey_ong_sellers_cmj_preprint.pdf.

Guy, Richard K. "The Strong Law of Small Numbers," *Amer. Math. Monthly*, 1988, **95**, no. 8, 697–712.

Hardy, G. H. *Ramanujan: Twelve Lectures on Subjects Suggested by His Life and Work*, Chelsea Publishing Company, New York, 1959.

Hardy, G. H., and E. M. Wright. *An Introduction to the Theory of Numbers*, 6th ed., Oxford University Press, Oxford, 2008. Revised by D. R. Heath-Brown and J. H. Silverman, with a foreword by Andrew Wiles.

Khare, Chandrashekhar, and Jean-Pierre Wintenberger. "Serre's Modularity Conjecture. I," *Invent. Math.*, 2009a, **178**, no. 3, 485–504.

———. "Serre's Modularity Conjecture. II," *Invent. Math.*, 2009b, **178**, no. 3, 505–586.

Klein, Jacob. *Greek Mathematical Thought and the Origin of Algebra*, Dover Publications, Inc., New York, 1992. Translated from the German and with notes by Eva Brann; reprint of the 1968 English translation.

Koblitz, Neal. *p-adic Numbers, p-adic Analysis, and Zeta-Functions*, 2nd ed., Graduate Texts in Mathematics, Vol. 58, Springer-Verlag, New York, 1984.

Mahler, K. "On the Fractional Parts of the Powers of a Rational Number II," *Mathematika*, 1957, **4**, 122–124.

Maor, Eli. *e: The Story of a Number*, Princeton University Press, Princeton, NJ, 2009.

Mazur, Barry. *Imagining Numbers: Particularly the Square Root of Minus Fifteen*, Farrar, Straus and Giroux, New York, 2003.

Nahin, Paul J. *Dr. Euler's Fabulous Formula: Cures Many Mathematical Ills*, Princeton University Press, Princeton, NJ, 2011.

Ono, Ken. 2015, http://www.mathcs.emory.edu/~ono/.

Pólya, George. *Mathematical Discovery: On Understanding, Learning, and Teaching Problem Solving*, John Wiley & Sons Inc., New York, 1981. Reprint in one volume, foreword by Peter Hilton, bibliography by Gerald Alexanderson, index by Jean Pedersen.

Ribet, Kenneth A., and William A. Stein. *Lectures on Modular Forms and Hecke Operators*, 2011, http://wstein.org/books/ribet-stein/main.pdf.

Ronan, Mark. *Symmetry and the Monster: One of the Greatest Quests of Mathematics*, Oxford University Press, Oxford, 2006.

Series, Caroline. "The Modular Surface and Continued Fractions," *J. London Math. Soc. (2)*, 1985, **31**, no. 1, 69–80.

Titchmarsh, E. C. *The Theory of the Riemann Zeta-function*," 2nd ed., The Clarendon Press, Oxford University Press, New York, 1986. Edited and with a preface by D. R. Heath-Brown.

Weil, André. *Foundations of Algebraic Geometry*, American Mathematical Society, Providence, RI, 1962.

Williams, G. T. "A New Method of Evaluating $\zeta(2n)$," *Amer. Math. Monthly*, 1953, **60**, 19–25.

# Index

absolutely convergent, 107
action, 151; discrete, 151
Ahlgren, Scott, 188
analytic continuation, 93
analytic function, 81
argument, 80
arithmetic series, 51
asymptotic formula, 187
Atkin, A.O.L., 212

Bernoulli; numbers, 59–69, 103,
    108, 170, 195; polynomials, 59–69
Bézout's identity, 12
binomial coefficient, 77
binomial series, 76–79
Birch, Bryan, 5, 218
Borcherds, Richard, 221
Boylan, Matthew, 188

$\mathbf{C}$, 80
Cauchy, Augustin-Louis, 8, 48
characteristic polynomial, 216
circle method, 41
Clozel, Laurent, 222
complex derivative, 81, 82
congruence group, 181; level, 182
Conway, John, 221
cusp, 152, 182
cusp form, 158, 182; normalized,
    202

$\Delta^*$, 99
$\Delta^0$, 82, 89, 99
Davenport, Harold, 40
Dedekind, Richard, 189
Deligne, Pierre, 196, 215
Dirichlet series, 113
disc of convergence, 88
discriminant, 173

$\eta$, 189
Eichler, Martin, 215
eigenbasis, 205
eigencuspform, 206
eigenvalue, 205
Eisenstein series, 168
elliptic function, 118
entire function, 94
equivalence relation, 15
Euler, Leonhard, 4, 7, 8, 23, 29, 104,
    107, 116, 190, 211

Fermat, Pierre de, 7, 23, 52
Fourier, Joseph, 191
fractional linear transformation, 145
Frobenius, Ferdinand, 215
fundamental domain, 130

$\Gamma$-function, 93, 211
$\Gamma(N)$, 181
$\Gamma_0(N)$, 181
$\Gamma_1(N)$, 181

$G(k)$, 38
$g(k)$, 37
generating function, 113
geodesic, 134
geometric series, 73–76, 81–83
Girard, Albert, 7
greatest common divisor, 12
Griess, Robert, 219

Hardy, G.H., 41, 187, 196
Harris, Michael, 222
Hasse, Helmut, 222
Hecke, Erich, 201
Hecke eigenform, 205; normalized, 206
Hilbert, David, 37
homomorphism, 220
hyperbolic plane, 142

imaginary part, 79
irreducible representations, 220

Jacobi, Carl, 118, 188, 190

Kant, Immanuel, 2
Klein, Felix, 134

Lagrange, Joseph-Louis, 34
Laurent series, 101
Lehner, Joseph, 212
linear combination, 161
Littlewood, J.E., 41

$M_k$, 159
McKay, John, 220
Mersenne, Marin, 7
modulus, 14
monodromy, 93
Mordell, Louis, 203

$n^s$, 94
newform, 201

norm, 80
Norton, Simon, 221

old form, 208
open set, 81
orbit, 151
Orwell, George, 3

partition, 111, 114
Pascal, Blaise, 52
Poisson summation, 191
polygonal numbers, 47
polynomial; monic, 104
prime, 14
Prime Number Theorem, 120
prime race, 26
punctured disc, 99

quadratic residue, 17
Quine, Willard Van Orman, 3

radius of convergence, 88
Ramanujan, Srinivasa, 173, 187, 188, 196
Rankin, Robert, 196
real part, 79
reciprocity law, 216
relatively prime, 12
residue; quadratic, 17
Riemann hypothesis, 94
Riemann surface, 93
Riemann $\zeta$-function, 94
Russell, Bertrand, 2

$S_k$, 159
Sato, Mikio, 221
Serre, Jean-Pierre, 215, 217
Shepherd-Baron, Nicholas, 222
Shimura, Goro, 215
subgroup, 148, 180
Swinnerton-Dyer, Peter, 5, 218

Tate, John, 221
Taylor, Richard, 200, 216, 218, 222

telescoping sum, 58
Type I prime, 24
Type III prime, 24

vector space, 161; basis, 163;
    dimension, 163; finite-dimensional,
    162; infinite-dimensional, 162
vector subspace, 161
Vinogradov, Ivan, 41

Waring, Edward, 37
weakly modular form, 219
Weil, André, 224
Wiles, Andrew, 200, 216, 218
Wilson's Theorem, 15
Wittgenstein, Ludwig, 3

**Z**, 12
Zeno, 74